Handbook of Aircraft Technology

Handbook of Aircraft Technology

Edited by **Natalie Spagner**

CLANRYE
INTERNATIONAL

New Jersey

Published by Clanrye International,
55 Van Reypen Street,
Jersey City, NJ 07306, USA
www.clanryeinternational.com

Handbook of Aircraft Technology
Edited by Natalie Spagner

International Standard Book Number: 978-1-63240-254-7 (Hardback)

Printed in the United States of America.

Contents

Preface

Aircraft technology is a dynamic field. This book provides insightful information regarding the advancements in aircraft technology. Specifically, it covers a broad range of topics in aircraft electrical systems, inspection and maintenance and other diversified topics. The authors are leading veterans in their fields. This book should appeal to both the students as well as researchers.

This book is the end result of constructive efforts and intensive research done by experts in this field. The aim of this book is to enlighten the readers with recent information in this area of research. The information provided in this profound book would serve as a valuable reference to students and researchers in this field.

At the end, I would like to thank all the authors for devoting their precious time and providing their valuable contribution to this book. I would also like to express my gratitude to my fellow colleagues who encouraged me throughout the process.

Editor

Part 1

Aircraft Electrical Systems

Power Generation and Distribution System for a More Electric Aircraft - A Review

Ahmed Abdel-Hafez
Shaqra University
Kingdom of Saudi Arabia

1. Introduction

More-Electric Aircraft (MEA) is the future trend in adopting single power type for driving the non-propulsive aircraft systems; i.e. is the electrical power. The MEA is anticipated to achieve numerous advantages such as optimising the aircraft performance and decreasing the operation and maintenance costs. Moreover, MEA reduces the emissions of air pollutant gases from aircrafts, which can contribute in signifcantly solving some of the problems of climate change. However, the MEA puts some challenges on the aircraft electrical system, both in the amount of the required power and the processing and management of this power. This chapter introduces the outline for MEA. It investigates possible topologies for the power system of the aircraft. The different electric power generation options are highlighted; while at the same time assessing the generator topologies. It also includes a general review of the power electronic interfacing circuits. Also, the key design requirements for an interfacing circuit are addressed. Finally, a glance at protection facilities for the aircraft power system is given.

2. More electric aircraft

Recently, the aircraft industry has achieved a tremendous progress both in civil and military sectors (AbdElhafez & Forsyth, 2008,2009; Cronin, 1990; Moir & Seabridge, 2001). For example some current commercial aircraft operate at weights of over 300 000 kg and have the ability to fly up to 16 000 km in non-stop journey at speed of 1000 km/h (AbdElhafez & Forsyth, 2009).

The non-propulsive aircraft systems are typically driven by a combination of different secondary power drives/subsystems such as hydraulic, pneumatic, electrical and mechanical (AbdElhafez & Forsyth, 2008,2009; Jones, 1999; Moir, 1999; Moir & Seabridge, 2001; Quigley, 1993). These powers subsystems are all soured from the aircraft main engine by different methods. For example, mechanical power is extracted from the engine by a driven shaft and distributed to a gearbox to drive lubrication pumps, fuel pumps, hydraulic pumps and electrical generators (AbdElhafez & Forsyth, 2009; Jones, 1999; Moir, 1999; Quigley, 1993). Pneumatic power is obtained by bleeding the compressor to drive turbine motors for the engine's starter subsystem, and wing anti-icing and Environmental Control Systems (ECS), while electrical power and hydraulic power subsystems are distributed throughout the aircraft for driving actuation systems such as flight control actuators, landing gear brakes, utility actuators, avionics, lighting, galleys, commercial loads and weapon systems (AbdElhafez & Forsyth, 2009, Howse, 2003; Jones, 1999; Moir, 1998, 1999; Quigley, 1993).

This combination had always been debated, because these systems had become rather complicated, and their interactions reduce the efficiency of the whole system. For example, a simple leak in pneumatic or hydraulic system jeopardises the journey by grounding the aircraft, and eventually causing inconvenient flight delays. The leak is usually difficult to locate and once located it cannot easily be handled (AbdElhafez & Forsyth, 2009; Cutts, 2002; Hoffman, 1985; Moir, 1998; Pearson, 1998; Rosero, et al, 2007; Weimer, 1993). Furthermore, from manufacturing point of view reducing the cost of ownership, increasing the profit and some anticipated future legislation regarding the climate changes demand radical changes to the entire aircraft, as it is no longer sufficient to optimise the current aircraft sub-systems and components individually to achieve these goals (AbdElhafez & Forsyth, 2009; Andrade, 1992; Cutts, 2002; Clyod, 1997; Emadi & Ehsani, 2000; Hoffman, 1985; Moir, 1998; Pearson, 1998; Ponton, 1998; Rosero, etal, 2007; Weimer, 1993).

The trend is using the electrical power for sourcing and distributing non-propulsive aircraft engine powers. This trend is defined as MEA. The MEA concept is utterly not a new concept, it has been investigated for several decades since W.W. II (Andrade, 1992; Cutts, 2002; Pearson, 1998; Ponton, 1998; Weimer, 1993). However, due to the lack of electric power generation capabilities and prohibitive volume of power conditioning equipments, the focus has been drifted into the conventional power types. Relatively, the recent technology breakthroughs in the field of power electronics systems, fault-tolerant electric machines, electro- hydrostatic actuators, electromechanical actuators, and fault-tolerant electrical power systems have renewed the interest in MEA (AbdElhafez & Forsyth, 2009; Andrade, 1992; Cutts, 2002; Clyod, 1997; Emadi & Ehsani, 2000; Hoffman, 1985; Moir, 1998; Pearson, 1998; Ponton, 1998; Rosero, etal, 2007; Weimer, 1993). A comparison between conventional aircraft subsystems and MEA subsystems is shown in Fig. 1 (AbdElhafez & Forsyth, 2009).

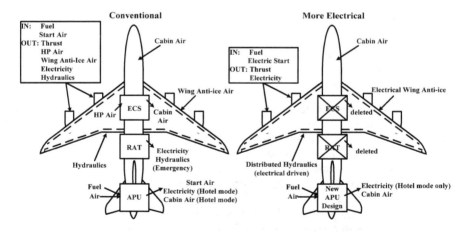

Fig. 1. Comparison between conventional systems aircraft and MEA systems (AbdElhafez & Forsyth, 2009).

The adoption of MEA in the future aircraft both in civil and military sectors will result in tremendous benefits such as:-

1. Removal of hydraulic systems, which are costly, labour-intensive, and susceptible to leakage and contamination problems, improves the aircraft reliability, vulnerability, and reduces complexity, redundancy, weight, installation and running cost (Cutts, 2002; Pearson, 1998; Ponton, 1998; Quigely, 1993; Weimer, 1993).
2. Deployment of electrical starting for the aero-engine through the engine starter/generator scheme eliminates the engine tower shaft and gears, power take-off shaft, accessory gearboxes and reduces engine starting power especially in the cold conditions and aircraft front area (Clyod, 1997; Emadi & Ehsani, 2000; Jones, 1999; Moir & Seabridge, 2001).
3. Utilization of the Advanced Magnetic Bearing (AMB) system, which could be integrated into the internal starter/generator for both the main engine and auxiliary power units, allows for oil-free, gear-free engine area (AbdElhafez & Forsyth, 2009; Andrade & Tenning, 1992a, 1992b; Hoffman et al., 1985; Jones, 1999; Moir & Seabridge, 2001).
4. In MEA, using a fan shaft generator that allowing emergency power extraction under windmill conditions removes the conventional inefficient single-shot ram air turbine, which increases the aircraft's reliability, and survivability under engine-failure conditions (AbdElhafez & Forsyth, 2009; Andrade & Tenning, 1992a, 1992b; Quigley, 1993).
5. Replacement of the engine-bleed system by electric motor-driven pumps reduces the complexity and the installation cost, and improves the efficiency (Jones, 1999).

In general, adopting MEA will revolutionise the aerospace industry completely, and significant improvements in terms of aircraft-empty weight, reconfigureability, fuel consumption, overall cost, maintainability, supportability, and system reliability will be achieved (AbdElhafez & Forsyth, 2009; Clyod, 1997; Cronin, 1990; Emadi & Ehsani, 2000; Hoffman et at., 1985; Moir, 21998, 1999, Weimer, 1993).

On the other hand, the MEA requires more demand on the aircraft electric power system in areas of power generation and handling, reliability, and fault tolerance. These entails innovations in power generation, processing, distribution and management systems (AbdElhafez & Forsyth, 2009; Clyod, 1997; Cronin, 1990; Emadi & Ehsani, 2000; Hoffman et at., 1985; Moir, 21998, 1999).

The proceeding sections briefly discuss a general overview of the electrical power distribution and management, generation and processing systems in MEA.

3. Distribution systems

The power distribution system of the most in-service civil aircrafts is composed of combined of AC and DC topologies. E.g., an AC supply of 115V/400Hz is used to power large loads as such as galleys, while the DC supply of 28V DC is used for avionics, flight control and battery-driven vital services.

Recently there is a trend for using only high voltage DC system for power distribution and management in MEA. A number of factors encouraged this trend (AbdElhafez & Forsyth,

2009; Cross et al., 2002; Hoffman, 1985; Jones, 1999; Glennon, 1998; Maldonado et al., 1996, 1997, 1999; Mallov et al., 2000; Quigely, 1993; Worth, 1990) :

1. Adopting the new generation options as variable frequency,
2. Recent advancements in the areas of interfacing circuits, control techniques and protection systems,
3. The advantages of the high voltage DC distribution system in reducing the weight, the size and the losses, while increasing the levels of the transmitted power.

Some values of the system voltage are presently under research. These values are: 270, 350 and 540V. The exact value, however, is determined by a number of factors such as, the capabilities of DC switchgear, the availability of the components and the risk of corona discharge at high altitude and reduced pressure (Brockschmidt, 1999).

Different topologies were suggested for implementing the distribution system in MEA (Cross et al., 2002; Hoffman, 1985; Glennon, 1998; Maldonado et al., 1996, 1997, 1999; Mallov et al., 2000; Worth, 1990). In the following four main candidates of these topologies are briefly reviewed, as follows :

1. Centralized Electrical Power Distribution System (CEPDS),
2. Semi-Distributed Electrical Power Distribution System (SDEPDS),
3. Advanced Electrical Power Distribution System (AEPDS),
4. Fault-Tolerant Electrical Power Distribution System (FTEPDS).

3.1 Centralized Electrical Power Distribution System (CEPDS)

CEPDS is a point-to-point radial power distribution system as shown in Figure 2. It has only one distribution centre. The generators supply this distribution centre. The electrical power is being processed and fed to the different electrical loads. The distribution centre is normally positioned in the avionics bay, Figure 2, where the voltage regulation is also located. In this system, each load is supplied individually from the power distribution centre (Cross et al., 2002; Worth et al., 1990). CEPDS has a number of advantages, such as :

1. The ease of maintenance, since all equipments are located in one place, i.e. avionics bay.
2. Decoupling between loads; thus the disturbance in a load is not transferred to the others.
3. Fault-tolerance, as the main buses are highly protected.

As stated   CEPDS may have significant advantages, however it also   has a number of disadvantages, such as:

1. CEPDS suffers from the difficulty of upgrading.
2. The faults in the distribution system affect probably all loads and disable the entire system.
3. CEPDS is cumbersome, expensive and unreliable, as each load has to be wired from the avionics bay.
4. Costly and bulky protection system has to be deployed to protect the distribution system.

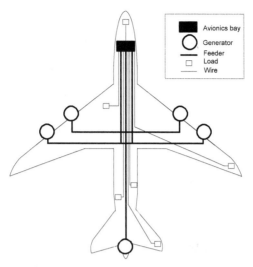

Fig. 2. Centralised Electrical Power Distribution System CEPDS for the MEA (AbdElhafez & Forsyth, 2009).

3.2 Semi-Distributed Electrical Power Distribution System (SDEPDS)

SDEPDS was proposed to overcome the problems of CEPDS (AbdElhafez & Forsyth; 2009; Cross et al., 2002; Hoffman, 1985; Glennon, 1998; Maldonado et al., 1996, 1997, 1999; Mallov et al., 2000; Worth, 1990) . The SDEPDS as shown in Figure 3 has a large number of Power Distribution Centres (PDCs). These centres are scaled versions of PDCs in CEPDS. The PDCs are distributed around the aircraft in such way to optimise the system volume, weight and reliability. They are located, Figure 3, close to load centres.

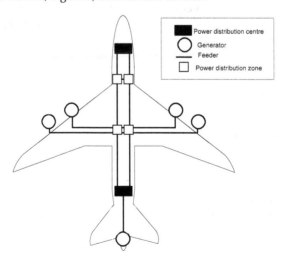

Fig. 3. Semi-Distributed Electrical Power Distribution System SDEPDS for the MEA (AbdElhafez & Forsyth, 2009)

SDEPDS has a number of advantages :

1. Elevated power quality and improved Electromagntic compatibility, due to the position of the distribution centres near to the loads,
2. High efficiency and cost effective, attributed to the deployment of electrical components with small weight/volume in PDCs,
3. Efficient and stable system operation, due to reduced losses/voltage drops across the distribution network.
4. High level of redundancy in primary power distribution path, due to the strategy of increasing and distributing the PDCs,
5. Simplicity and flexibility of upgrading.

On the other hand, the close coupling between the loads in SDEPDS may reduce the reliability, as faults/ disturbances in a load can propagate to nearby loads. Moreover, extra equipments are required to perform the monitoring and control of the distributed PDCs.

3.3 Advanced Electrical Power Distribution System (AEPDS)

AEPDS is a flexible, fault-tolerant system controlled by a redundant microprocessor system. This system is developed to replace the conventionally centralized and semi-distributed systems.

AEPDS as shown in Figure 4, is highly protected. The electrical power from the generators, Auxiliary Power Unit (APU), battery and ground sources is supplied to the primary power distribution, where the Contactor Control Units (CCU) and high power contactors are located. The primary power distribution centre performs a number of tasks: voltage/frequency regulation, damping oscillation and transient and controlling the flow of the reactive power.

The aircraft loads are supplied via the Relay Switching Units (RSU). Each RSU is controlled and monitored by a Remote Terminal (RT) unit. The AEPDS is controlled by either one of the two redundant Electrical load Management Units (ELMU). The ELMU interact and exchange data/control strategies with the RTs through a quad redundant data bus (Mollov et al., 2002; Worth, 1990) .

The AEPDS has improved performance than CEPDS and SDEPDS. This is attributed for the following (Worth, 1990):

1. AEPDS reduces the aircraft life cycle cost, as the system reconfiguration in case of aircraft modification/upgrade can easily be accommodated.
2. AEPDS can detect deviant conditions of current/voltage and provide instantaneous load shut-off.
3. A major reduction in the weight and wiring in the AEPDS is achieved due to the elimination of circuit breaker panels from the flight deck stands.
4. AEPDS is fault-tolerant distribution system.

The AEPDS has the disadvantage of concentrating the distribution and the management of power supplied by the generating units/sources into a single unit; therefore a fault in this unit may interrupt the whole system operation.

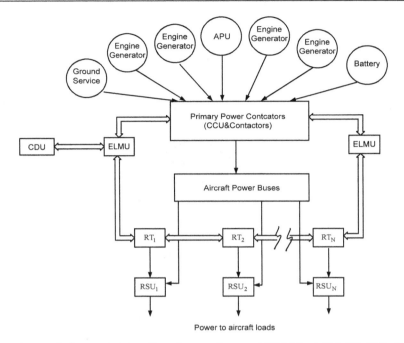

Fig. 4. Advanced Electrical Power Distribution System AEPDS for MEA (AbdElhafez & Forsyth, 2009).

3.4 Faulted-Tolerant Electrical Power Distribution System (FTEPDS)

FTEPDS is adequatly protected. A typical FTEPDS for a two-engine aircraft is shown in Figure 5. The system is composed of two switch matrices, six multi-purpose converter, six generators and different loads. The source and load switch matrices could be implemented by using mechanical or solid-state switches. However, the latter has the advantages of controllability, fast response and high efficiency (Cross et al., 2002; Hoffman et al., 1985; Glennon, 1998; Maldonado et al., 1996, 1997) over the former.

FTEPDS is a mixed distribution system; the AC power from generators and airport grid are connected to source switch matrix, while 270V DC system is interfaced with the converters. The bi-directional power flow in the generators indicates that system allows integral starter/generator operation, where the generator initally acts as a motor to start the jet engine; then it operates as generator to supply the aircraft electrical system. Also 270V DC system has a bi-directional power flow; this is to charge the batteries and other energy storage units during normal flight conditions. However, during faults and disturbances the DC system injects power to stabilize the aircraft distribution system.

FTEPDS enjoys the following advantages:

1. The ability to start the aircraft engine by generator/starter scheme,
2. High redundancy,
3. Fault-tolerant, the ability of the system to continue functioning even under an engine failure,

However FEEPDS has a serious drawback; a fault in source/load switch matrices may interrupt the operation of the entire system.

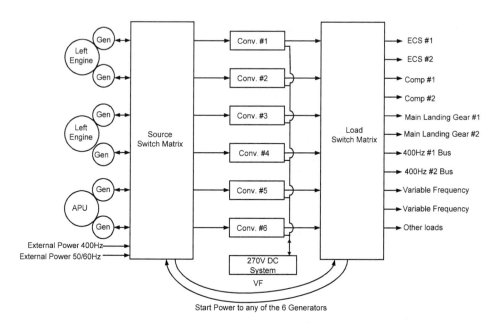

Fig. 5. Fault-tolerant Electrical Power Distribution System FTEPDS for MEA (AbdElhafez & Forsyth, 2009).

4. Electric power generation in MEA

Since its advent, generated electrical power uilization has been rising rapidly. The growth of electrical power generation/application in aircrafts is shown in Figure 6. The quadratic growth is attributed to the increase aircraft system loading such as : galley and In-Flight Entertainment (IFE) systems.

MEA recently is one of the major driving force in electric generation in aircrafts (AbdElhafez & Forsyth, 2009; Andrade, 1992; Bansal et al., 2003, 2005; Howse, 2003; Jones, 1999; Quigely, 1993; Mellor et al., 2005; Moir & Seabridge, 2001; Moir, 1999; Raimondi et al., 2002). Not only are aircraft electrical system power levels growing, but the diversity of the power generation types is increasing as well.

4.1 Schemes of power generation

The various in-service and prospect schemes of electrical power generation are shown in Figure 7 (AbdElhafez & Forsyth, 2009; Cossar, 2004)

Examples of civil/military aircraft and the corresponding generation scheme are given in Table 1.

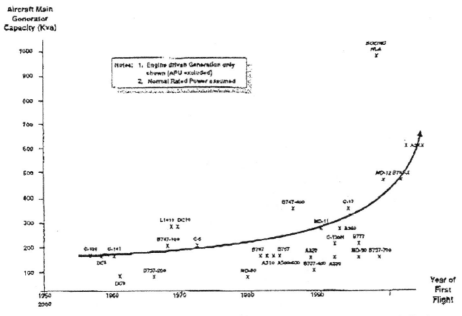

Fig. 6. Growth of generated electrical power in aircraft since the first flight (AbdElhafez & Forsyth, 2009).

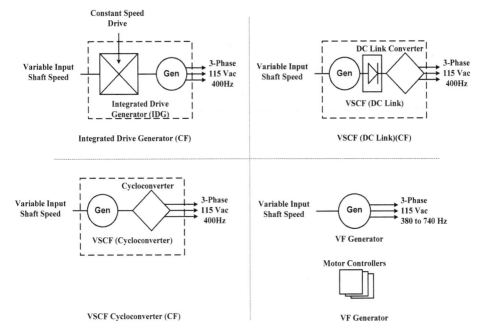

Fig. 7. Aircraft Electrical Power Generation Options (AbdElhafez & Forsyth, 2009).

Generation scheme	Civil		Military	
	Aircraft	Rating (kVA)	Aircraft	Rating (kVA)
CF(IDG)	B777	2x120		
	A340	4x90		
	B373NG	2x90		
	MD-12	4x120		
	B747-X	4x120		
	B717	2x40		
	B767-400	2x40		
	Do728	2x40		
VSCF (Cycloconverters)			F-18E/F	2x60/65
VSCF(DC-link)	B77(backup)	2x20	C145	2x120
	MD-90	2x75		
VF	Global Ex	4x50	Boeing JSF	2x50
	Horizon	2x20/25		
	A3xx	4x150		
270VDC			F-22 Raptor	2x70
			X-35A/B/C	2x50

Table 1. Civil/Military aircraft and electrical power generation techniques (AbdElhafez & Forsyth, 2009).

A brief review of the different generation techniques is given below where the focus is on the merits/demerits of each.

4.1.1 Constant frequency

The constant Frequency (CF), three-phase 115V/400Hz scheme is the most common electric power generation option. This scheme is in-service in most civil aircrafts as shown in Table 1. The CF is alternatively termed Integrated Drive Generator (IDG).

In CF system, the generator is attached to the engine through unreliable and cumbersome mechanical gearbox. This gearbox is essential to ensure that the generator speed is constant irrespective of the engine speed and aircraft status. The frequency f of the generated power is related to generator speed N by,

$$f = \frac{PN}{120} \tag{1}$$

where f is output frequency in cycle/sec(Hz); N is generator speed in revolution per minutes (rpm) and P is the number of magnetic poles. Maintaining generator speed N constant ensures that output frequency remains fixed; however the CF has a number of disadvantages (AbdElhafez & Forsyth, 2009; Cossar, 2004; Howse, 2003; Jones, 1999; Quigely, 1993; Moir, 1999; Raimondi et al., 2002):

1. The interfacing mechanical gear box is unreliable, inefficient and costly, which reduces the overall system efficiency.

2.   The system has to be examined for every flight, increasing the operational costs.
3.   CF could not allow internal starting for the aero-engine by integral starter/generator scheme.

4.1.2 DC-link system

Variable Speed Constant Frequency (VSCF) DC-link system is now the preferred option for most new military aircraft and some commercial aircraft, Table 1. The generator in this scheme, Figure 7, is attached directly to the engine, thus according to (1) the output frequency will vary with engine speed. The engine speed is subjected to wide variation during the normal course of flight, and so does the frequency; therefore interfacing circuits are required to change the generator output power into usable form.

The output of the generator is supplied to diode rectifiers, which converts the variable frequency AC power into DC form. Then three-phase inverters are used to convert the DC power into three-phase 115V/400Hz AC type. This is the typical form of VSCF DC-link system. However, recently several topologies were reported. These new topologies produce improved performance regarding harmonics, reactive power flow and system stability. Moreover, the range of VSCF DC-link system has been widened due to the recent advancements in field of high power electronic switches. VSCF DC-link option is generally characterised by simplicity and reliability (AbdElhafez & Forsyth, 2009; Hoffman et al., 1985; Ferriera, 1995; Moir, 1999; Quigley, 1993; Olaiya, &. Buchan, 1999; Ying shing & Lin, 1995).

4.1.3 Cycloconverters

Variable Speed Constant Frequency (VSCF) Cycloconverters as shown in Figure 2 convert directly the variable frequency AC input power into AC form with fixed frequency and amplitude, three-phase 115V/400Hz (AbdElhafez & Forsyth, 2009; Cloyd, 1997; Cronin, 1990; Emad & Ehasni, 2000; Howse, 2003, Jones, 1999; Moir & Seabridge, 2001). The output frequency is lower than the input frequency; thus, making it possible for the generator to be attached to the engine with a fixed turns ratio gearbox. In the typical form of cycloconverters, three bidirectional switches interface each generator phase with the corresponding supply phase.

The VSCF cycloconverters are more efficient than CF and VSCF DC-link; however they require sophisticated control. The power generation efficiency of the cycloconverters increases as the power factor decrease, which would be beneficial if this technique is applied to motor loads with significant lagging power factors (AbdElhafez & Forsyth, 2009).

4.1.4 Wild frequency

Variable Frequency (VF), commonly known as wild frequency, is the most recent electric power generation contender. In VF approach, the generator is attached directly to the engine shaft. This method is commonly termed embedded generation (Raimondi et al., 2002). Generator direct allocation in the engine shafts de-rates power take-off shaft and the associated gearbox, which reduce their size and weight and increase the reliability. However, a number of implications will arise, in case of embedding one or more electrical machines within the core of the engine:

1. Accommodation of the embedded generators requires revision of the design of the engine components from their current state, which may change the components structure and probably the profile of the airflow through the engine.
2. The heat loss within the generator places a significant burden on the engine oil cooling system, requiring additional or alternative heat exchange.
3. If the generator rotor is only supported through main engine bearings, the small air gap requirement of the generator may lead to obligatory stiffening of the engine structure. The latter being nessary to ensure that rotor and stator do not come into contact under high acceleration
4. Transmitting high levels of electrical power to and from the core of the engine would require significant alterations in the supporting engine core structure relative to the engine pylon (Raimondi et al., 2002).

In VF, variations in engine speed would manifest directly into the output frequency as shown from (1) and Figure 2. The promising features of VF are the small size, weight, volume, and cost as compared with other aircraft electrical power generation options. Also VF offers a very cost-effective source of power for the galley loads, which consumes a lot of on-board power. However VF may pose significant risk at higher power levels, particularly with high power motor loads. Furthermore, the cost of motor controllers required due to the variation in the supply frequency, need to be taken into consideration when assessing VF (AbdElhafez & Forsyth, 2009; Cronin, 2005; Elbuluk & Kankam, 1997; Hoffman, 1999; ; Moir, 1998, Pearson, 1998; Weimer, 1993).

4.2 Generator topologies

The anticipated increase in electrical power generation requirements on MEA suggests that high power generators should be attached directly to the engine, mounted on the engine shaft and used for the engine start in Integral Starter/Generator (IS/G) scheme . The harsh operating conditions and the high ambient temperatures push most materials close to or even beyond their limits, requiring more innovations in materials, processes and thermal management systems design.

Consequently, Induction, Switched Reluctance, Synchronous and Permanent Magnet machine types (Hoffman et al., 1985; Mollov et al., 2000; Cross, 2002) have been considered for application in MEA due to their robust features.

4.2.1 Induction generator

Induction Generators (IGs) are characterized by their robustness, reduced cost and ability to withstand harsh environment. However, the IG requires complex power electronics and is considered unlikely to have the power density of the other machines (Khatounian et al.,2003; Ying & Lin, 1995; Bansal et. al, 2003, 2005).

4.2.2 Synchronous generator

The current generator technology employed on most commercial and military aircraft is the three-stage wound field synchronous generator (Hoffman, 1985). This machine is reliable and inherently safe; as the field excitation can be removed, de-energising the machine.

Therefore, the rating of the three-stage synchronous generator has increased over the years reaching to 150KVA (Hoffman, 1985) on the Airbus A380. The synchronous machine has the ability to absorb/generate reactive power, which enhances the stability of the aircraft power system. However, this machine requires external DC excitation, which unfortunately decreases the reliability and the efficiency.

4.2.3 Switched reluctance generator

The Switched Reluctance (SR) machine has a very simple robust structure, and can operate over a wide speed range. The three-phase type has a salient rotor similar to salient pole synchronous machine. The stator consists of three phases; each phase is interfaced with the DC supply through two pairs of anti-parallel switch-diode combination. Thus, the SR machine is inherently fault-tolerant. However the machine has the severe disadvantage of producing high acoustic noise and torque ripples (Mitcham & Cullenm, 2002, 2005; Pollock & Chi-Yao, 1997; Trainer & Cullen, 2005; Skvarenina et al., 1996,1997).

4.2.4 Permanent generator

The Permanent Magnet (PM) generator has a number of favourable characteristics (AbdElhafez & Forsyth, 2009; Argile, 2008; Bianchi, 2003; Jack et al., 1996; Pollock & Chi-Yao, 1997; Mecrow et al., 1996; Mitcham & Cullenm, 2002, 2005):

1. Ease of cooling, as the PM generator theoretically has almost zero rotor losses.
2. High efficiency compared to other machine types.
3. High volumetric and gravimetric power density.
4. High pole number with reduced length of stator end windings.
5. Self excitation at all times.

However, conventional PM machines are claimed to have inferior fault tolerance compared with SR machines (Argile, 2008; Mecrow et al., 1996; White, 1996). Conventional PM generators are intolerant to elevated temperatures. Furthermore, PM generators require power converters with high VA rating to cater for a wide speed range of operation (AbdElhafez & Forsyth, 2009; Bianchi, 2003 ;Jack et al., 1996;Mecrow et al., 1996; Mitcham & Cullenm, 2002, 2005). Therefore, a different implementation is mandatory in PM machine technology if they are to be used in aero-engines.

The fault-tolerant PM machines are one solution and offer high levels of redundancy and fault tolerance (Argile, 2008; Ho et al.,1988; Mitcham & Grum, 1998; Mellor, at al.,2005). These machines are designed with a high number of phases, such that the machine can continue to deliver a satisfactory level of torque/power after a fault in one or more phases. Furthermore, each phase has minimal electrical, magnetic, and thermal impact upon the others (Argile, 2008; Jack et al., 1996; Jones & Drager, 1997;Mecrow et al., 1996; Mitcham & Cullenm, 2002, 2005; White, 1996). This is realised by:

1. The number of magnetic poles in the machine being similar to the stator slot number; each phase winding can be placed in a single slot, which is thermally isolated from the other phases (AbdElhafez, 2008; Adefajo, 2008; Jones & Drager, 1997;Mecrow et al., 1996; Mitcham & Cullenm, 2002).
2. The stator coils being wound around alternate teeth, which provides physical and magnetic isolation between the phases (AbdElhafez, 2008; Jones & Drager, 1997).

3. Each phase being attached to a separate single-phase power converter, which achieves the electrical isolation (AbdElhafez, 2008; Adefajo, 2008; Jack et al., 1996; Jones & Drager, 1997;Mecrow et al., 1996; Mitcham & Cullenm, 2002, 2005).
4. The machine synchronous reactance per phase is typically 1.0 p.u., limiting the short-circuit fault current to no greater than the rated phase current (AbdElhafez, 2008; Jack et al., 1996; Jones & Drager, 1997;Mecrow et al., 1996; Mitcham & Cullenm, 2002, 2005).

4.3 Integrated generation

MEA as mentioned, suggests innovative strategies for optimizing the aircraft performance and reducing the installation and operational costs, such as IS/G and emergency power generation schemes.

4.3.1 Integral starter/generator

Commonly, jet engines are externally started by pneumatic power from a ground cart. This reduces the system reliability and increases maintenance and running cost. A move toward internal starting for the engine is adopted in MEA.

The jet engine has two shafts: High Pressure (HP) and Low Pressure (LP) shafts. The main generator is usually attached to the HP shaft . The trend is to use that generator as the prime mover to start the engine. Once the engine is started, the generator returns to its default operation, generator. The prime mover (starter) is powered from the aircraft system, which during this stage is supplied from energy storage devices. ISG scheme has a number of advantages (AbdElhafez & Forsyth, 2009; Ganev, 2006; Elbuluk & Kankam, 1997; Ferreira, 1995; Skvarenina, 1996, 1997) :

1. Improves the aircraft reconfigureability by eliminating the arrangement used previously for ground starting.
2. Allows the adoption of All Electric Aircraft (AEA)
3. Uses AMB system that results in reliable robust and compact engine.
4. Reduces the operational and maintenance cost, which boosts the air traffic industry

Different machine topologies are suggested for IS/G scheme; however the SR and fault-tolerant PM machines are most reliable. These machines do not require external excitation or sophisticated control techniques. Also, they are either inherently or artificially fault-tolerant.

4.3.2 Emergency power generation

The level of the emergency power is expected to grow significantly for future aircrafts, due to rising demands of critical aircraft loads/services. Currently, the emergency power is sourced from generators coupled to a Ram Air Turbine (RAT). This scheme is deployed only under emergency conditions, and suffers from serious drawbacks such as (AbdElhafez et al., 2006a, 2006b, 2008; Adefajo, 2008; Bianchi, 2003) :

1. It is expensive to develop, install and maintain.
2. It is unpopular with the airliners.
3. The integrity of such a 'one-shot' system is always subject to some doubt.

The proposal is to utilize the windmill effect of the aero-engine fan, which is driven from the LP shaft, for emergency power generation. While, the fan is normally rotating, the heath of the emergency generation system is continuously monitored and backup power will be immediately available following a main generator failure. Also the stored inertial energy of the engine is significant and could be recovered as another source of emergency power (AbdElhafez & Forsyth, 2008, 2009; Ganev, 2006).

Different machine topologies are competing for LP emergency generators. Trade-off studies were conducted to identify the most suitable machine technology. Due to the difficulty of the location, reliability is paramount and it is clear that a brushless machine format is required. The harsh operating environment particularly extremely high ambient temperatures, pushes many common materials, e.g. permanent magnet materials and insulation materials close to or beyond their operating limits. Consequenclty, cooling or alternative materials and process would be required (AbdElhafez & Forsyth, 2008, 2009; Mitcham &. Grum, 1998) .

Machine efficiency is another crucial issue, since dissipated heat needs to be absorbed by the engine cooling system. Currently, the generator loss is absorbed by the engine oil system and this is in turn mainly cooled by the fuel entering the engine. This restricts the amount of heat that can be dissipated without introducing an alternative cooling method.

Some key requirements, assisting in the choice of LP generator type are list below (AbdElhafez & Forsyth, 2008, 2009):

1. The machine operates only as a generator, drive torque is not allowed.
2. The machine is subject to a harsh operating environmental conditions (specifically high temperature), with limited access for maintenance.
3. Power must be generated over a very wide speed range (approximately 12:1) with an output voltage compatible with the aircraft DC-distribution system voltage 350 V dc.
4. The machine is fault tolerant, such that it continues to run even if there is a fault on one or two phases without significantly degrading the output power.

Also the operating speed range, weight and volume constraints are important parameters that affect the choice of machine type.

Several brushless machine types seem to have the required ruggedness and hence the capability of operation in such environment. These include: IG, SR and PM machines (AbdElhafez & Forsyth, 2008, 2009; Mitcham &. Grum, 1998).

5. Interfacing circuits

There are many occasions within the aircraft industry where it is required to convert the electrical power from one level/form to another level/form, resulting in a wide range of Power Electronics Circuits (PECs) such as AC/DC, DC/DC, DC/AC and matrix converters (AbdElhafez & Forsyth, 2009; Chivite-Zabalza, 2004; Cutts, 2002; Lawless & Clark, 1997; Matheson, &. Karimi, 2002; Moir & Seabridge, 2001; Singh et. al, 2008). There are general requirements, which PEC should satisfy:

1. PEC should have reduced weight and volumetric dimension.
2. PEC should be fault-tolerant, which implies its ability to continue functioning under abnormal conditions without much loss in its output power or degradation of its performance.

3. PEC should be efficient and have the ability for operation in harsh conditions such as high temperature and low maintenance.
4. PEC should emit minimum levels of harmonic and Electromagnetic Interference (EMC).
5. PEC could be easily upgraded and computerized.

Innovation in the area of power electronics components is required to enable realisation of MEA. Wide-Band Gap (WBG) High-Temperature Electronics (THE) is an example of these developments. The devices manufactured from WBG-THE are capable of operating at both higher temperatures (600 ^{0}C) (Reinhardt & Marciniak, 1996) and higher efficiencies compared to Si-based devices (-55 ^{0}C to 125 ^{0}C). A number of advantages are expected to be realized from employing WBG-THE devices (AbdElhafez et al., 2006, 2008, Howse, 2003; Gong et al., 2003; Lawless & Clark, 1997; Matheson, &. Karimi, 2002; Moir & Seabridge, 1998, 2001; Trainer & Cullen, 2005):

1. Eliminating/reducing of ECS required for cooling flight control electronics and other critical PECs
2. Reducing the engine control system weight and volumetric dimension
3. Improving the system reliability by using a distributed processing architecture
4. Optimizing the aircraft system and reducing the installation and running cost
5. Improving system fault-tolerance and redundancy

Another main challenge for PECs in the aircraft is passive electrical component size, as the current components are heavy and bulky, especially for the high power level expected in the MEA. However, the on-going research in the design and fabrication of the passive components for MEA gives some optimistic results. For example, some advanced polymer insulation materials such as Eymyd, L-30N, and Upilex S (AbdElhafez & Forsyth, 2009; Cutts, 2002; Lawless & Clark, 1997; Moir & Seabridge, 2001) have the ability to operate over a wide temperature range (-269 ^{0}C to 300 ^{0}C). Also these materials can withstand the environmental conditions such as humidity, ultraviolet radiation, basic solution and solvent at high altitudes (AbdElhafez & Forsyth, 2009; Lawless & Clark, 1997). The ceramic capacitor is a good example, which offers remarkable advantages in volumetric density compared to other capacitor technology (Lawless & Clark, 1997).

	SSCB	Conventional
Mechanism	The breaker consists of bidirectional switches that allow current flow in both directions. The gating signal of the switches are blocked to inhabit the faulty current	Commonly an isolating air gap is developed in the path of the fault current. A upon disconnection, an arc is created. Depending on the arc distinguishing methodology the breaker is termed.
Response time	Very small	Long
Power rating	Small	Medium to high
Volumetric/weight	Compact/small	Bulky/heavy
Cost	Expensive	Cheap
Functionality	Multi-task, they perform current monitoring and status reporting	They should be instructed to be opened

Table 2. Comparison between SSCB and conventional breakers.

6. Protection system

The distribution system of aircraft is adequatly protected; different types of Circuit Breakers (CBs) are utilized. Thus includes the conventional and power electronics based. The conventional CBs include air, SF6, and oil, while the Solid-State Circuit Breakers (SSCBs) represent the power electronics based breakers (AbdElhafez & Forsyth, 2009; Jones, 1999; Moir & Seabridge, 2001). A comparison between SSCB and a generic conventional CB is given in Table 2 above.

7. References

AbdElhafez, A. (2008). *Active Rectifier Control for Multi-Phase Fault-Tolerant Generators*. PhD desertion, University of Manchester, UK.

AbdEl-Hafez, A.; Cross, A.; Forsyth, A.; Mitcham, A.; Trainer, D.& Cullen, J. (2006). Fault Tolerant Starter-Generator Converter Optimisation ",Patent Application Rolls-Royce, Ed. UK, 2006.

AbdEl-hafez, A.; Cross, A.; Forsyth, A.;Trainer, D.& Cullen, J. (2006). Single-Phase Active Rectifier Selection for Fault Tolerant Machine, *in 3rd IET International Conference on Power Electronics, Machines and Drives, PEMD* 2006, pp. 435-439, April 2006.

AbdElHafez, A & Forsyth, A. J. (2009) . A Review of More-Electric Aircraft, *Proceedings of The 13rd international conference on Aerospace Science and Aviation Technology conference* ,ASAT-13. Cairo, Egypt, May 26-28, 2009.

AbdEl-Hafez, A.; Todd, R.; Forsyth, A.& Long, S. (2008). Single-Phase Controller Design for a Fault Tolerant Permanent Magnet Generator, in *IEEE Vehicle Power and Propulsion Conference, VPPC 2008*, pp. 250-257, September 2008.

Adefajo, O.; Barnes, M.; Smith, A.; Long, S.; Trainer, D.; AbdEl-hafez, A.; & Forsyth, A. (2008). Voltage Control On An Uninhabited Autonomous Vehicle Electrical Distribution System," in *The 4th IET International Conference on Power Electronics, Machines and Drives, PEMD 2008*, pp. 676-680, April 2-4, 2008.

Andrade, L. & Tenning, C.(1992). Design of the Boeing 777 Electric System, *IEEE National Aerospace and Electronics Conference,*.pp.1281 - 1290, May 18-22, 1992.

Andrade, L. & Tenning, C.(1992). Design of Boeing 777 electric system, *IEEE Aerospace and Electronic Systems Magazine*, Vol. 7, (1992) pp. 4-11.

Argile, R.; Mecrow, B.; Atkinson, D.; Jack, A.& Sangha, P. (2008). reliability analysis of fault tolerant drive topologies, *the Proceeding of The 4th IET International Conference on Power Electronics, Machines and Drives,PEMD 2008*, pp 11-15, April 2-4, 2008.

Bansal, R.; Bhatti, T.& Kothari, D.(2003). Bibliography on the application of induction generators in nonconventional energy systems," *IEEE Transaction on Energy Conversion*, Vol. 18, (September 2003), pp. 433-439.

Bansal, R. (2005). Three-phase self-excited induction generators: an overview, *IEEE Transaction on Energy Conversion*, Vol. 20, (20005), pp. 292-299.

Bianchi, N.; Bolognani, S.; Zigliotto, M. & Zordan, M. (2003). Innovative remedial strategies for inverter faults in IPM synchronous motor drives, *IEEE Transaction on Energy Conversion,*, Vol. 18, (June 2003), pp. 306-314.

Brock, A. & Schmidt, T. (1999). Electrical environments in aerospace applications, *in Proceedings of International Conference Electric Machines and Drives, IEMD '99*, pp. 719-721, May 9-12, 1999.

Chivite-Zabalza, F.; Forsyth, A. & Trainer, D. (2004). Analysis and practical evaluation of an 18-pulse rectifier for aerospace applications, *in Second International Conference on Power*

Electronics, Machines and Drives, PEMD 2004, Vol.1, pp. 338-343, March-31 April-2, 2004.

Cloyd, J. (1997). A status of the United States Air Force's More Electric Aircraft initiative," in *Proceedings of the 32nd Intersociety Energy Conversion Engineering Conference, IECEC-97,* Vol.1, pp. 681-686, July-27 August-1 1997.

Cross, M.; Forsyth, A & Mason, G. (2002) . Modelling and simulation strategies for the electric system of large passenger aircraft," in *SAE 2002 conference,* pp. 450-459, 2002.

Cutts, S. (2002). A collaborative approach to the More Electric Aircraft, in *Proceedings of International Conference on Power Electronics, Machines and Drives,PEMD 2002,* pp. 223-228, April 16-18, 2002.

Cossar, C.& Sawata, T. (2004). Microprocessor controlled DC power supply for the generator control unit of a future aircraft generator with a wide operating speed range, in *Second International Conference on Power Electronics, Machines and Drives, PEMD 2004,* Vol.2, pp. 458-463, March 31, April-2, 2004.

Cronin, M. (1990). The all-electric aircraft, *IEE Review,* Vol. 36, (September 1990), pp. 309-311.

Elbuluk, M.& Kankam, P. (1997). Potential starter/generator technologies for future aerospace applications, *IEEE Aerospace and Electronic Systems Magazine,* Vol. 12, (May 1997),pp. 24-31.

Emadi, K &. Ehsani, M. (2000). Aircraft power systems: technology, state of the art, and future trends, *IEEE Aerospace and Electronic Systems Magazine,* Vol. 15, (January 2000), pp. 28-32.

Ferreira, C.; Jones, S.; Heglund, W.& Jones, W. (1995). Detailed design of a 30-kW switched reluctance starter/generator system for a gas turbine engine application, *IEEE Transactions on Industry Applications,* Vol. 31, (May/June 1995). pp. 553-561.

Ganev, E. (2006). High-Reactance Permanent Magnet Machine for High-Performance Power Generation Systems" *SAE Power Systems Conference,* pp. 247-253, November, 2006.

Glennon, T. (1998). Fault tolerant generating and distribution system architecture," in *IEE Colloquium on All Electric Aircraft,* (June 1998), pp. 1-4.

Gong, G.; Drofenik, U.& Kolar, J. (2003). 12-pulse rectifier for more electric aircraft applications, in *IEEE International Conference on Industrial Technology,* Vol.2, pp. 1096-1101, December 10-12, 2003.

Hoffman, A; Hansen, A. Beach, R.; Plencner, R.; Dengler, R.; Jefferies, K. & Frye, R. (1985) Advanced secondary power system for transport aircraft, *NASA Technical Paper* 2463, (May 1985) http://ntrs.nasa.gov/archive/nasa/19850020632_1985020632.pdf

Hoffman, A.; Hansen, I.; Beach, R.; Plencner, R.;Dengler, R.; Jefferies, K. & Frye, J. (1985) "Advanced secondary power system for transport aircraft," in *IEE Colloquium on All Electric Aircraft,* (June 1995), pp. 1-4.

Ho, T.; Bayles, R. & Sieger, E. (1988). Aircraft VSCF generator expert system," *IEEE Aerospace and Electronic Systems Magazine,* Vol. 3, (April 1988), pp. 6-13.

Howse, M. (2003). All electric aircraft, *Power Engineer Journal,* Vol. 17, (2003) pp. 35-37.

Jack, A.; Mecrow, B. and Haylock, J. (1996). A comparative study of permanent magnet and switched reluctance motors for high-performance fault-tolerant applications, *IEEE Transactions on Industry Applications,* Vol. 32, (July/August 1996), pp. 889-895.

Jones, R. (1999). The More Electric Aircraft: the past and the future?, in *IEE Colloquium on Electrical Machines and Systems for the More Electric Aircraft,*(November 1999), pp. 1-4.

Jones, S. & Drager, B. (1997). Sensorless switched reluctance starter/generator performance,*IEEE Industry Applications Magazine,* Vol. 3, (1997), pp. 33-38.

Khatounian, F.; Monmasson, E.; Berthereau, F.; Delaleau, E.& Louis, J. (2003). Control of a doubly fed induction generator for aircraft application, in *Proceedings of 29th Annual*

Conference of the IEEE Industrial Electronics Society, IECON '03, Vol.3, (November 2003), pp. 2711-2716.

Lawless, W.& Clark, C. (1997). Energy storage at 77 K in multilayer ceramic capacitors, *IEEE Aerospace and Electronic Systems Magazine,* , Vol. 12, (August 1997), pp. 32-35.

Maldonado, M. & Korba, G. (1999). Power management and distribution system for a more-electric aircraft (MADMEL)," *IEEE Aerospace and Electronic Systems Magazine,* Vol. 14, (1999), pp. 3-8.

Maldonado, M.; Shah, N.; Cleek, K.; Walia, P. & Korba, G. (1996). Power management and distribution system for a more-electric aircraft (MADMEL)-program status," *in Proceedings of the 32nd Intersociety Energy Conversion Engineering Conference, IECEC-97.* Vol. 1, pp. 148-153, 1996.

Maldonado, M.; Shah, N.; Cleek, K.; Walia, P. & Korba, G. (1997). "Power Management and Distribution System for a More-Electric Aircraft (MADMEL)-program status," *in Proceedings of the 33nd Intersociety Energy Conversion Engineering Conference, IECEC-97.* Vol. 1, pp. 274-279. 1997.

Matheson, E.; Karimi, K. (2002). Power Quality Specification Development for More Electric Airplane Architectures, in *SAE International Conference,* Vol. 2, pp. 343-347, 2002.

Mecrow, B.; Jack, A.; Haylock, J. & Coles, J. (1996). Fault-tolerant permanent magnet machine drives, *IEE Electric Power Applications,* Vol. 143, (November 1996), pp. 437-442.

Mellor, P.; Burrow, S.; Sawata, T.& Holme, M. (2005). A wide-speed-range hybrid variable-reluctance/permanent-magnet generator for future embedded aircraft generation systems, *IEEE Transactions on Industry Applications,* Vol. 41, (Marc-April 2005), pp. 551-556.

Mitcham, A. ; Antonopoulos , G. & Cullen, J. (2002). Favourable slot and pole number combinations for fault-tolerant PM machines, *in Proceedings of IEE Electric Power Applications,* Vol. 151, (September 2004), pp. 520-525.

Mitcham, A. & Grum, N. (1998). An integrated LP shaft generator for the more electric aircraft. *in IEE Colloquium on All Electric Aircraft,* (June 1998), pp. 1-9.

Mitcham, A. & Cullen, J. (2005). Permanent Magnet Modular Machines: New design Philosophy, in *Electrical Drive Systems for the More Electric Aircraft one-Day Seminar,* pp. 1-8, 2005.

Mitcham, A. & Cullen, J. (2002). Permanent magnet generator options for the More Electric Aircraft, *in Proceeding of International Conference on Power Electronics, Machines and Drives, PEMD* 2002, pp. 241-245, April 16-18, 2002.

Moir, I. (1999). .More-electric aircraft-system considerations, *in IEE Colloquium on Electrical Machines and Systems for the More Electric Aircraft,* (1999), pp. 1-9.

Moir, I. & Seabridge, A. (2001). *Aircraft systems : mechanical, electrical, and avionics subsystems integration,* London press, London, UK

Moir, I. (1998). .The all-electric aircraft-major challenges," *in IEE Colloquium on All Electric Aircraft,* (June 1998), pp. 1-6.

Mollov, S.; Forsyth, A. & Bailey, M. (2000). System modelling of advanced electric power distribution architecture for large aircraft," *SAE Transaction* (2000), pp. 904-913.

Olaiya, M.& Buchan, N. (1999) "High power variable frequency generator for large civil aircraft," in *IEE Colloquium on Electrical Machines and Systems for the More Electric Aircraft,* (November 1999), pp. 1-4.

Pearson, W. (1998). The more electric/all electric aircraft-a military fast jet perspective, *in IEE Colloquium on All Electric Aircraft* (June 1998), pp. 1-7.

Ponton, A. & at al (1998). "Rolls-Royce Market Outlook 1998-2017," *Rolls-Royce Publication No TS22388* (1998).

Pollock C. & Chi-Yao, W. (1997). Acoustic noise cancellation techniques for switched reluctance drives," *IEEE Transactions on Industry Applications*, Vol. 33, (March/April 1997), pp. 477-484.

Provost, M. (2002). The More Electric Aero-engine: a general overview from an engine manufacturer, *in Proceedings of International Conference on Power Electronics, Machines and Drives, PEMD 2002*, pp. 246-251, April 16-18 , 2002.

Quigley, R. (1993). .More Electric Aircraft, *in Proceedings of Eighth Annual Applied Power Electronics Conference and Exposition, APEC '93*, pp. 906-911, March 1993.

Raimondi, C.; Sawata, T.; Holme, M.; Barton, A.; White, G.; Coles, J.; Mellor, P. & Sidell, N (2002). Aircraft embedded generation systems, *in Proceeding of International Conference on Power Electronics, Machines and Drives*, PEMD 2002, pp. 217-222, April 16-18, 2002.

Reinhardt, K.& Marciniak, M. (1996). Wide-band gap power electronics for the More Electric Aircraft, *in Proceedings of the 31st Intersociety Energy Conversion Engineering Conference, IECEC 96.*, pp. 127 – 132, August 11-16, 1996.

Richter, E. & Ferreira, C. (1995). Performance evaluation of a 250 kW switched reluctance starter generator, *in Thirtieth IAS Annual Meeting IEEE Industry Applications Conference, IAS '95.*, Vol.1, pp. 434-440, October 8-12, 1995.

Rosero, J; Ortega, J.; Aldabas, E. & Romeral, L. (2007) Moving towards a more electric aircraft, *IEEE Aerospace and Electronic Systems Magazine*, Vol. 22, (2007), pp. 3-9.

Shing, Y.& Lin, C. (1995). A prototype induction generator VSCF system for aircraft, *in International IEEE/IAS Conference on Industrial Automation and Control: Emerging Technologies*, pp. 148-155, May 22-27, 1995.

Singh, B.; Gairola, S.; Singh, N.; Chandra, A.& Al-Haddad, K. (2008). Multiples AC-DC Converters for Improving Power Quality: A Review, *IEEE Transactions on Power Electronics*, Vol. 23, (January 2008.), pp. 260-281.

Skvarenina, T.; Pekarek, S.; Wasynczuk, O.; Krause, P.; Thibodeaux, R. & and Weimer, J. (1997), Simulation of a switched reluctance, More Electric Aircraft power system using a graphical user interface, *in Proceedings of the 32nd Intersociety Energy Conversion Engineering Conference, IECEC-97*, 1997, Vol.1, pp. 580-584, July-27 August - 1, 1997.

Skvarenina, T.; Wasynczuk, O.; Krause, P.; Zon, W.; Thibodeaux, R.& Weimer, J. (1996). Simulation and analysis of a switched reluctance generator/More Electric Aircraft power system, *in Proceedings of the 31st Intersociety Energy Conversion Engineering Conference, IECEC 96.*, Vol.1, pp. 143-147, August 11-16, 1996.

Trainer, D. & Cullen, J. (2005). Active Rectifier for Fault Tolerant Machine Application, Derby, internal memorandum, February 24, 2005.

White, R. & Miles, M. (1996). Principles of fault tolerance, *the Proceeding of Eighth Annual Applied Power Electronics Conference and Exposition, APEC '96*, Vol.1, pp. 18-25, 1996.

Weimer, J. (1993). Electrical power technology for the more electric aircraft, *in Proceedings of 12th AIAA/IEEE Digital Avionics Systems Conference, DASC 1993*, pp. 445-450, October 25-28, 1993.

Welchko, B.; Lipo, T.; Jahns, T. & Schulz, S.(2004). Fault tolerant three-phase AC motor drive topologies: a comparison of features, cost, and limitations, *IEEE Transactions on Power Electronics, Vol. 19*, (July 2004), pp. 1108-1116.

Worth, F.; Forker, V.; Cronin, M. (1990). Advanced Electrical System (AES)," *in Aerospace and Electronics Conference*, pp. 400 - 403, 1990.

2

Power Electronics Application for More Electric Aircraft

Mohamad Hussien Taha
Hariri Canadian University
Lebanon

1. Introduction

In the competitive world of airline economics, where low cost carriers are driving dawn profit margins on airline seat miles, techniques for reducing the direct operating costs of aircraft are in great demand. In effort to meet this demand, the aircraft manufacturing industry is placing greater emphasis on the use of technology, which can influence maintenance costs and fuel usage. (Faleiro, 2005)

There is a general move in the aerospace industry to increase the amount of electrically powered equipments on future aircraft. This trend is referred to as the "More Electric Aircraft". It assumes using electrical energy instead of hydraulic, pneumatic and mechanical means to power virtually all aircraft subsystem including flight control actuation, environmental control system and utility function. The concept offers advantages of reduced overall aircraft weight, reduced need for ground support equipment and maintenance and increased reliability (Taha,2007,Wiemer,1999).

Many aircraft power systems are now operating with a variable frequency over a typical range of 360 Hz to 800 Hz.

Distribution voltages for an aircraft system can be classified as:

a. Nominal 115/200 V rms and 230/400 V rms ac, both one phase and three phase, over variable frequency range.
b. Nominal 14, 28 and 42 V DC.
c. High DC voltage which could be suitable for use with an electric actuator (or other) aircraft loads.

This chapter presents studies, analysis and simulation results for a boost and buck converters at variable input frequency using vector control scheme. The design poses significant challenges due to the supply frequency variation and requires many features such as:

1. The supply current to the converter must have a low harmonic contents to minimize its impact on the aircraft variable frequency electrical system.
2. A high input power factor must be achieved to minimize reactive power requirements.
3. Power density must be maximized for minimum size and weight.

2. Boost converter for aircraft application

A three phase boost converter which is shown in fig. 1 with six steps PWM provide DC output and sinusoidal input current with no low frequency harmonic. However the switching frequency harmonics contained in the input currents must be suppressed by the input filter. Referring to fig.1 after the output capacitor has charged up via the diodes to a voltage equals to $1.73V_{pk}$, the diodes are all reverse biased. Turning one of the MOSFETs in each of the three phases will cause the inductor current to increase. Assume the input voltage V_a is positive, if S_2 is turned on, the inductor current increases through the diode D_4 or D_6 and the magnetic energy is stored in the inductor. Since the diodes D_1, D_3 and D_5 are reverse biased, the output capacitor C_{dc} provides the power to the load. When S_2 is turned off, the stored energy in the inductor and the AC source are transferred to C_{dc} and the load via the diodes. When the AC voltage is negative, S_1 is turned on and the inductor current increases through the diode D_3 or D_5. The same operation modes are involved for phase B and phase C (Taha., 2008; Habetler.,1993). Fig.2 and fig. 3 show different operating modes.

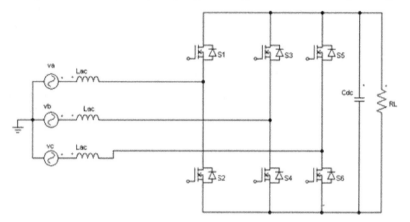

Fig. 1. Boost converter.

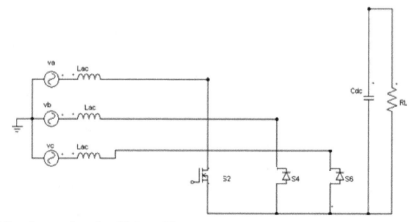

Fig. 2. Boost converter when V_a is positive.

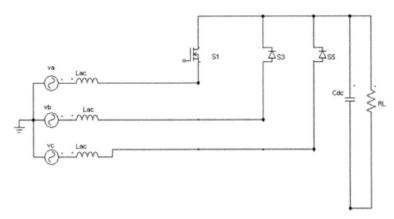

Fig. 3. Boost converter when V_a is negative.

3. Buck converter for aircraft application

The buck 3-phase/dc converter is a controlled current circuit which relies on pulse width modulation of a constant current to achieve low distortion. As shown in fig. 4. The circuit consists of 3 power MOSFETs and 12 diodes, an AC side filter and DC side filter.

The AC side input and DC side output filters are standard second order low pass L-C filters. For the input filter, the carrier frequency has to be considerably higher than the filter resonance frequency in order to avoid resonance effects and ensure carrier attenuation. The Ac side filter is arranged to bypass the commutating energy when the MOSFETs are turning off and to absorb the harmonic for the high frequency switching. At the DC side, the inductor is used to maintain a constant current, this inductor can be relatively small since the ripple frequency will be related to the switching frequency. The magnitude and the phase of the input current can be controlled and hence the power transfer that occurs between the AC and DC sides can also be controlled. (Green et al., 1997).

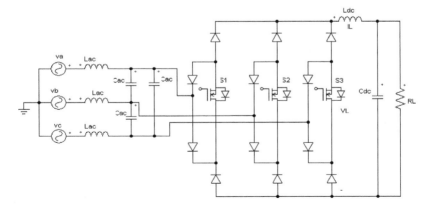

Fig. 4. Buck converter.

The input phase voltages V_a, V_b, V_c and the input currents I_a, I_b, I_c are assumed to be sinusoidal of equal magnitude and symmetrical

$$V_a = V_{pk} \sin(\omega t) \quad I_a = I_{pk} \sin(\omega t + \varphi) \tag{1}$$

$$V_b = V_{pk} \sin(\omega t - 2\pi/3) \quad I_b = I_{pk} \sin(\omega t - 2\pi/3 + \varphi) \tag{2}$$

$$V_c = V_{pk} \sin(\omega t + 2\pi/3) \quad I_c = I_{pk} \sin(\omega t + 2\pi/3 + \varphi) \tag{3}$$

Figure (5) shows 60 degrees of two sine waveforms.

$$T_a = TM \sin(\omega t + \varphi) \tag{4}$$

$$T_b = TM \sin(\omega t + 2\pi/3 + \varphi) \tag{5}$$

The freewheeling time T_f is equal to:

$$T_f = T - T_a - T_b \tag{6}$$

Where: φ is the displacement angle,
Vpk is the peak phase voltage,.
M modulation index.
T is the PWM switching period.

The general operation of the system is as follows: The switching of the devices is divided into six equal intervals of the 360 degrees main cycle. The waveforms repeat a similar pattern at each interval. At any time during the switching interval, only two converter legs are modulated independently and the third leg is always on. There are some time intervals that only one device is only on, thus providing a freewheeling for the DC current since at this time the energy stored on the DC inductor feeds the load.

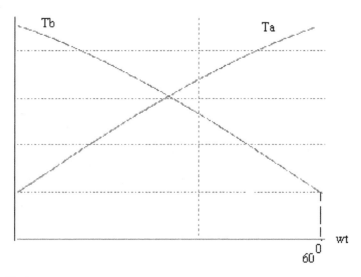

Fig. 5. Two 60° sine waves.

Mode 1

With S_2 on and S_1 is modulated by reference T_a, current flows in phase (a) and phase (b). $I_a>0$ and $I_b<0$, The bridge output voltage V_L is connected to main line supply V_{ab} which opposed by V_{DC} is applied across the inductor. Current I_L increases in the inductor.

$$I_L = I_a = -I_b \tag{7}$$

$$V_L = V_{ab} \tag{8}$$

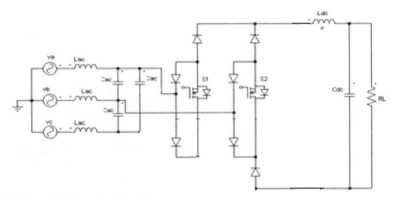

Fig. 6. Mode 1 equivalent circuit.

Mode 2

With S_2 on and S_3 is modulated by T_b, current flows in phase (c) and phase (b). $I_c>0$ and $I_b<0$. Line voltage V_{cb} opposed by V_{DC} is applied across the inductor. Again the inductor current increases.

$$I_L = I_c = -I_b \tag{9}$$

$$V_L = V_{cb} \tag{10}$$

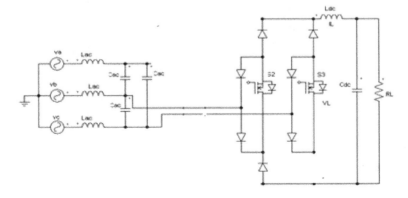

Fig. 7. Mode 2 equivalent circuit.

Mode 3

In this mode, only one MOSFET is on (S_2), the inductor current freewheels and the converter is disconnected from the mains and the DC voltage is zero.

$$V_L = 0 \tag{11}$$

Therefore the average voltage V_L over one switching period T is:

$$V_L = [(V_{ab} \times T_a) + (V_{cb} \times T_b)] / T \tag{12}$$

Where:

$$V_{ab} = V_a - V_b = 1.5 V_{pk} \sin(\omega t) + 0.866 V_{pk} \cos(\omega t) \tag{13}$$

$$V_{cb} = V_c - V_b = 1.73 V_{pk} \cos(\omega t) \tag{14}$$

By substituting equations 8, 10, 12 and 13 into equation 12 yeilds:

$$V_L = V_{DC} = [(V_{ab} \times T_a) + (V_{cb} \times T_b)] / T = 1.5 \, M \, V_{pk} \cos\varphi \tag{15}$$

By assuming an ideal power converter in which the power losses are negligible, the power nput is then equal to power output, and by assuming $\cos\varphi = 1$, The DC output voltage can be defined as :

$$V_{DC} = 1.5 \, M \, V_{pk} \tag{16}$$

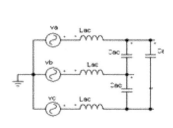

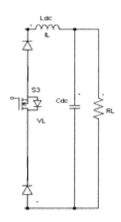

Fig. 8. Mode 3 equivalent circuit.

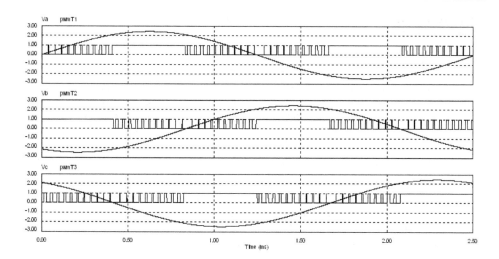

Fig. 9. PWM for the buck converter.

4. Adaptive reactive power control using boost converter

In comparison to the operating frequencies of land based power systems, which is normally 50-60 Hz, the operation of aircraft power systems at these relatively high frequencies can present some technical difficulties. One area of importance is associated with the impedance of the potentially long cables (which may run along part of the wing and a large proportion of the fuselage). These cables connect electrical loads, such as electric actuators for aircraft flight surfaces, to the AC supply, or "point of regulation" (POR). In large modern aircraft, the cables can be in excess of 200 ft and contribute impedance which is dependent on the cable's inductance and resistance. The inductive reactance, X_L, is proportional to the operating frequency of the power system and is given by $X_L = 2\pi fL$, where f is the operating frequency and L is the inductance of the cable and therefore the reactance changes with operating frequency. (Taha M, Trainer R D 2004). As the connected load draws a current, the cable develops a voltage drop due to its impedance which is out of phase with respect to the voltage at the POR and has two detrimental effects:

The voltage at the load is reduced below the regulated voltage at the point of regulation which is usually at the generator output.

The power factor of the load seen at the point of regulation reduces (even for a purely resistive load).

The voltage drop across the cable is clearly disadvantageous. The voltage drop may be tolerated and the connected loads have to be correspondingly down rated for the lower received voltage. Alternatively the voltage drop across the length of the cable is not allowed to exceed a threshold (typically 4 V) and it is necessary to provide cables that are both large and heavy such that their resistance remains low. Clearly space and weight are at a premium in aerospace applications. There can be significant weight saving if smaller, high resistance cables are used, particularly where low duty cycle, pulsed loads like electric actuators are supplied. The detrimental effects of such cables may be offset if the system designer uses the high inductive reactance present at the higher operating frequencies to affect voltage boost and power factor correction.

The simplest type of compensation for this type of problem is to connect a set of 3-phase capacitors (star or delta) at the point of connection of the load, in a similar way to ac motor-start capacitors. The capacitors can be used as a generator of reactive power but the beneficial effects are limited since the capacitive compensation is mainly controlled by the voltage magnitude and system frequency rather than the requirements of the load. Having noted the limitations of connecting shunt capacitors, there may be some applications where this type of compensation is applicable.

There is growing interest in the use of advanced power electronic circuits for aerospace loads, particularly in the motor-drives associated with electric actuators. The main two classes of converters currently being considered are active rectifiers and direct ac-ac frequency changer circuits (e.g. Matrix converters). Both types of converter can be made to operate with leading, lagging or unity power factor by suitable control of the semi-conductor switching elements.

The current view in the aerospace industry appears to be that the operation of these converters should be limited to unity power factor and little (or no) work has been carried out to explore the true system level benefits of variable power factor operation.

Fig.10 shows a basic circuit diagram for an electric actuator load incorporating an advanced power electronic converter with power factor control. It is clear that by controlling the power factor of the converter (shown leading), the effects of cable inductance can be eliminated so that the load as seen from the POR becomes unity power factor. Other operating power factors may be desirable in order to optimize the operation of the overall power system, including the generator loading.

Because the effects are proportional to the load current flowing through the cable and the system frequency, the reactive power compensation provided by the converter also needs to be variable.

The voltage magnitude at the load can be made the same as that at the POR. It could be beneficial in some applications to boost the input voltage by increasing the capacitive compensation provided by the power electronic converter.

The main benefit of using the advanced power electronic converter as a source of reactive power is to reduce (or eliminate) the voltage drop down the connecting cable. This gives us the possibility to use high impedance cables with benefits of reduced conductor diameter and significantly lower weight.

In order to understand the benefits of reactive power control, it is convenient to consider the flow of real and reactive current separately as shown in figure 10. Superposition can then be used to assess the net effect of both forms of current flow.

Therefore:

$$i = i_p + ji_q = I\left(\cos\theta_1 + j\sin\theta_1\right) \qquad (17)$$

Where

$$\theta_1 = \tan^{-1}(i_q/i_p) \tag{18}$$

$$E = V + (i_p + ji_q)R + (i_p + ji_q)jX_L \tag{19}$$

$$E = V + i_p R - i_q X_L + j(i_q R + i_p X_L) \tag{20}$$

$$E = E ((\cos\theta_2 + j\sin\theta_2) \tag{21}$$

Where

$$\theta_2 = \tan^{-1}((i_q R + i_p X_L)/ (V + i_p R - i_q X_L)) \tag{22}$$

For unity power factor θ_1 should equals θ_2.
Therefore:

$$(i_q/i_p) = (i_q R + i_p X_L)/ (V + i_p R - i_q X_L) \tag{23}$$

$$i_q = ((V/ X_L) \pm ((V/ X_L)^2 - 4 i_p^2)^{1/2})/2 \tag{24}$$

In a practical system, ip could take the form of a current demand and iq would be a separate reactive current demand that is made to vary as a function of ip and X_L (frequency dependant). R and X_L are cable dependant parameters.

Referring to Fig. 10, the inputs here are system frequency and load current, the output is Q demand, which is an input to the power electronic converter. The parameters of the cable are stored and used within the electronic circuitry to calculate the required compensation for the system under consideration.

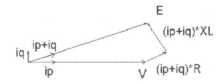

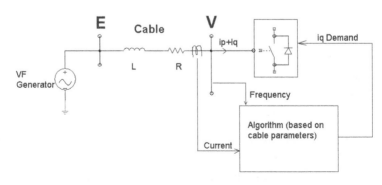

Fig. 10. System performance for reactive power compensation.

5. DQ vector control for the converters

In the DQ vector control strategy the instantaneous 3 phase voltages and currents are transferred to a 2-axis reference frame system which rotates at the angular frequency of the supply. This has the effect of transforming the three phase AC quantities (representing rotating volt and current phasors in the stationary co-ordinate frame) into DC quantities in the synchronously rotating frame (Taha et al.,2002, Taha,. 2008). If the D axis is chosen to be aligned with the voltage phasor, the D and Q axis current components represent the active and reactive components respectively. Fig. 11 shows the schematic of the DQ control scheme implemented in the input converter.

The proposed control scheme consists of two parts:

1.   An outer voltage controller.
2.   An inner current controller.

The outer voltage controller regulates the DC link voltage. The error signal is used as input for the PI voltage controller this provides a reference to the D current of the inner current controller. The Q current reference is set to zero to give unity power factor. A PI inner current control is used to determine the demand of the stationary DQ voltage values (Taha M & Trainer R D 2004; Kazmierkoski et al., 1991).

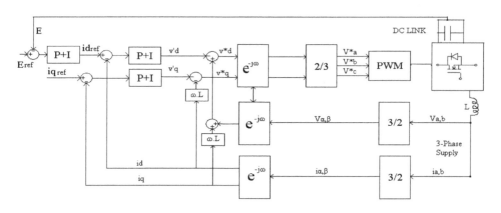

Fig. 11. DQ Control lock diagram.

Each gain in the controller affects the system characteristics differently. Settling time, steady state error and system stability are affected by the amount of the proportional gain. Selecting a large gain attains faster system response, but cost of large overshoot and longer settling time. Application of the integral feedback drives the steady state error to zero. The integral term increases as the sum of the steady state error increases causing the error to eventually be zero. However it can cause overshoot and ringing.

Selection of the two gain constants is critical in providing fast system response with good system characteristics.

The general formulas for DQ transformations are given as follows. We assume that the three-phase source voltages v_a, v_b and v_c are balanced and sinusoidal with an angular frequency ω.

The components of the input voltage phasor along the axes of a stationary orthogonal reference frame (α, β) are given by:

$$v_\alpha = v_a \tag{25}$$

$$v_\beta = \frac{1}{\sqrt{3}} \, (2 \, v_b + v_a) \tag{26}$$

The input voltage can then be transformed to a rotating reference frame DQ chosen with the D axis aligned with the voltage phasor. The voltage components are given by:

$$v_d = v_\alpha \cos \omega t - v_\beta . \sin \omega t \tag{27}$$

$$v_q = v_\alpha \sin \omega t + v_\beta \cos \omega t \tag{28}$$

The same transformations are applied to the phase currents.

$$i_d = i_\alpha \cos \omega t - i_\beta . \sin \omega t \tag{29}$$

$$i_q = i_\alpha \sin \omega t + i_\beta \cos \omega t \tag{30}$$

Let v_{a1}, v_{b1} and v_{c1} be the fundamental voltages per phase at the input of the converter.

$$v_a = Ri_a + L \, di_a/dt + v_{a1} \tag{31}$$

$$v_b = Ri_b + L \, di_b/dt + v_{b1} \tag{32}$$

$$v_c = Ri_c + L \, di_c/dt + v_{c1} \tag{33}$$

where L is the value of input line inductance and R is its resistance of the inductor.

Taking the steady state DQ transformation for the inductor, the input voltage to the converter in the DQ reference frame is given by:

$$v_d = Ri_d + L.di_d/dt - \omega Li_q + v_{d1} \tag{34}$$

$$v_q = Ri_q + L.di_q/dt + \omega Li_d + v_{q1} \tag{35}$$

The active and reactive powers are given by:

$$P = v_d.i_d + v_q.i_q \tag{36}$$

$$Q = v_d.i_q - v_q.i_d \tag{37}$$

Inverse DQ transformations then need to be applied to provide the three phase modulating waves (v_{aref}, v_{bref} and v_{cref}) for the PWM generation.

The main advantages of the DQ control are :

1. Direct control the active and reactive power.
2. Fast dynamics of current control loops.

The PWM generator based on a regular asymmetric PWM strategy.

Voltage Control

The DC side may be modelled by a capacitor C, representing the smoothing capacitors, and a resistor R, representing the load. This is shown in Figure 12.

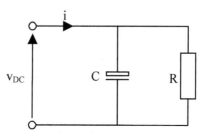

Fig. 12. Schematic of dc voltage link.

The linearised model for the DC side is given by the open-loop transfer function relating the DC link voltage to the supply current:

$$G(s) = \frac{v_{DC}(s)}{i(s)} = \frac{R}{1 + RCs} \tag{38}$$

Applying the PI controller illustrated, i(s) is given by:

$$i(s) = \left(K_p + \frac{K_i}{s}\right)(v_{REF} - v_{DC}) \tag{39}$$

Thus, the closed-loop transfer function is given by:

$$\frac{v_{DC}(s)}{v_{REF}(s)} = \frac{\left(K_p s + K_i\right)/C}{s^2 + \dfrac{s\left(1 + RK_p\right)}{RC} + \dfrac{K_i}{C}} \tag{40}$$

To give a damped response, the poles of the system should be placed along the real axis in the s-domain, i.e. at $s = -\omega_1$ and $s = -\omega_2$, giving the transfer function:

$$\frac{v_{DC}(s)}{v_{REF}(s)} = \frac{\left(K_p s + K_i\right)/C}{s^2 + s\left(\omega_1 + \omega_2\right) + \omega_1 \omega_2} \tag{41}$$

By equating the coefficients of the denominators of the above equations, the proportional and integral gains are:

$$K_p = C\left(\omega_1 + \omega_2\right) - 1/R \tag{42}$$

$$K_i = C\omega_1\omega_2 \tag{43}$$

The zero of the transfer function is where:

$$\left(K_p s + K_i\right)\big/C = \left(\left(C(\omega_1 + \omega_2) - 1/R\right)s + C\omega_1\omega_2\right)\big/C = 0 \tag{44}$$

Therefore:

$$s = \frac{-C\omega_1\omega_2}{C(\omega_1 + \omega_2) - 1/R} \tag{45}$$

An approach to the controller design is to locate this zero to coincide with one of the poles, say at $s = -\omega_1$, so as to cancel its effect. This gives:

$$\omega_1 = 1/RC \tag{46}$$

The second pole can then be placed at any desired location, to give the desired bandwidth. This gives the proportional and integral gains:

$$K_p = C\omega_2 \tag{47}$$

$$K_i = \omega_2/R \tag{48}$$

Current Control for boost converter

In this case the system is the line from the generator to the input converter, which may be modelled by an inductor in series with a resistor. The generator e.m.f. is assumed to have no dynamic effect, and so is represented as a short circuit. The system schematic is shown in Fig. 13.

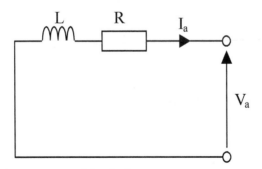

Fig. 13. Schematic of current control for the boost converter.

The phase current is given by:

$$i_a = -\frac{v_a}{R} - \frac{L}{R}\frac{di_a}{dt} \tag{49}$$

The open loop transfer function relating the phase current to the phase voltage is, therefore:

$$\frac{i_a}{v_a} = -\frac{1}{R + Ls} \tag{50}$$

From this simple transfer function, it would appear that the PI controller proposed would suffice, driving the steady-state current error to zero and allowing the behaviour and bandwidth, (i.e. the positions of the poles, of the closed loop system) to be fully determined by choosing the proportional and integral gains. The D-axis and Q-axis currents are compared to their respective demanded values and the error is applied to individual PI controllers to give voltage demands referred to the D-axis and Q-axis. With the feed-forward and dc-coupling terms, the transfer functions of the systems being controlled are:

$$\frac{i_d(s)}{v_d{}'(s)} = \frac{1}{Ls + R} \tag{51}$$

$$\frac{i_q(s)}{v_q{}'(s)} = \frac{1}{Ls + R} \tag{52}$$

Again, these are first-order equations and similar to the voltage control loop, the PI controllers will drive the steady-state error to zero and enable the behaviour and bandwidth of the closed-loop system to be determined by placing the poles appropriately.

Current Control for buck converter

The idea of controlling the current of the AC side LC filter has been proposed as a way of suppressing the excitation of the resonance of this filter. In steady state and in the absence of distortion there are no current components to excite the resonance because the resonant frequency will have been chosen to fall between the fundamental and the switching frequency. During the transient, the resonance of the filter can be damped by choosing the characteristics impedance to match the resistance and the inductance.

In this case the system is the line from the generator to the input converter, which may be modeled by an inductor in series with a resistor and capacitor.. The system schematic is shown in Figure 14.

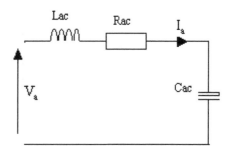

Fig. 14. Schematic of current control for the buck converter.

The open loop transfer function relating the phase current to the phase voltage is, therefore:

$$G(s) = \frac{1}{s^2 L_{ac}C_{ac} + sR_{ac}C_{ac} + 1} \tag{53}$$

From this transfer function it would be appear that the PI controller proposed would suffice, driving the steady state error to zero and allowing the behaviour and bandwidth (position of the poles) of closed loop system to be fully determined by choosing the proportional and integral gains. The procedure for this is the same as that described above for voltage control.

6. Hardware design

All of the converters components had to be selected so that normal service maintenance would ensure the retention of their specified characteristics through the full range of operational and environmental conditions likely to be encountered through the life of the aircraft, or support facility, in which they are installed (Taha M 1999).

6.1 Capacitors

The choice of capacitors is very important for aerospace industry. Wet aluminum electrolytic capacitors are not suitable due to their limited operating temperature range and hence limited life. Equivalent series resistance is also a problem for these and other types of electrolytic capacitor and therefore alternative technologies, such as ceramic or plastic, are recommended.

Ceramic capacitors have a good lifetime, low series resistance and they work in high temperature conditions. On the other hand for a rating of a few hundred volts this type of capacitor has a very small value per unit volume and are only available in units of up to 20uF. The size and weight for this converter are very important. Therefore care was taken to choose the optimal value of the DC capacitor

6.2 Magnetic components

Another important factor is the design of the magnetic components. In order to achieve a small air gap, minimum winding turns, minimum eddy current losses and small inductor size, the inductor should be designed to operate at the maximum possible flux density. Also, care should be taken to ensure that the filter inductors do not reach a saturated state during the overload condition. As the cores saturate, the inductance falls and the THD rises.

7. Simulation results for boost and buck converter

The power conversion in the boost or buck converter is exclusively performed in switched mode. Operation in the switch mode ensures that the efficiency of the power conversion is high. The switching losses of the devices increase with the switching frequency and this should preferably be high in order have small THD therefore choosing the switching frequency poses significant challenges due to:

1. Supply frequency Variation (360 to 800 Hz).

For the boost converter the simulation carried out with a fixed switching frequency. However, for the buck converter. one of the method could be used is a variable switching frequency which depend on the input frequency. Trade off between the values of the filters and the switching frequency have been studies, in order to maintain the THD within the required value at different input frequency. Another method is to use the same switching frequency for different input frequency, here the highest input frequency should be considered.

The parameter values used for the simulation are shown in table 1. Fig. 15 to fig. 18 show, input AC voltage and current and Dc output voltage.

Boost converter	Buck converter
RMS phase voltage = 115 V	RMS phase voltage = 115 V
DC voltage setting = 400 V	DC voltage setting = 42 V
AC Input Filter = L_{ac}= 100uH	AC Input Filter = L_{ac}= 150uH; C_{ac} = 1μF
Dc Output filter C_{dc} = 50μF	Dc Output filter = L_{dc} = 1mH ; C_{dc} = 50μF
Load = 10 Ω	Load = 0.5 Ω
Switching freq for 800 input freq =20000 Hz	Switching freq for 800 input freq =33600 Hz
Switching freq for 360 input freq =20000 Hz	Switching freq for 360 input freq =23760 Hz

Table 1. Simulation parameters.

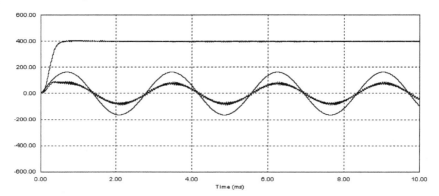

Fig. 15. Boost converter simulation results at 360 Hz input frequency.

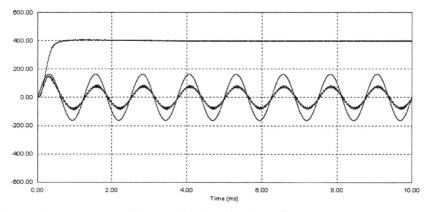

Fig. 16. Boost converter simulation results at 800 Hz input frequency.

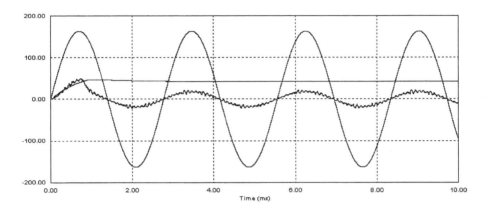

Fig. 17. Buck converter simulation results at 360 Hz input frequency.

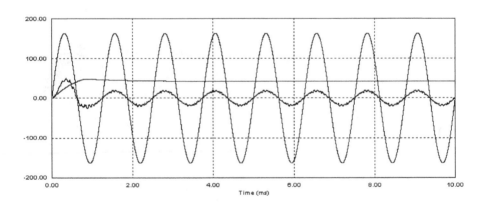

Fig. 18. Buck converter simulation results at 800 Hz input frequency.

8. Simulation results for the adaptive power control

Simulation has been done at 16 kW approximately. Fig. 19 shows the "per phase" parameter values used. Fig. 20 and fig. 21 show results for 360 Hz, the voltage drops to 107.5 V at the input filter of the converter. To compensate for the voltage drop across the cable, q (reactive demand) has been set this gave leading power factor. Fig. 21 shows that the voltage Va3 increase to 111.3 V at 0.9 PF.

Fig. 22 and Fig. 23 show results for 800 Hz, the voltage drops to 106 V at the input filter of the converter. Fig. 23 shows that the voltage Va3 increase to 1113 V at 0.9 PF.

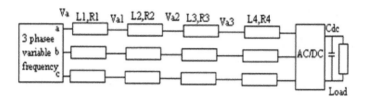

V_{a1} is the point of regulation voltage

V_{a3} is the point of connection of the load

L_1 = 25uH,R_1= 0.015 ohms is the generator inductance and resistance.

L_2 = 10uH,R_2= 0.01 ohms is the cable inductance and resistance from the generator to contactor.

L_3 = 20uH,R_2= 0.1 ohms is the cable inductance and resistance from the contactor to the load.

L_4 = 100uH,R_4= 0.1 ohms is the inductance and resistance of the load converter input filter.

Fig. 19. Single phase parameters for the adaptive power control.

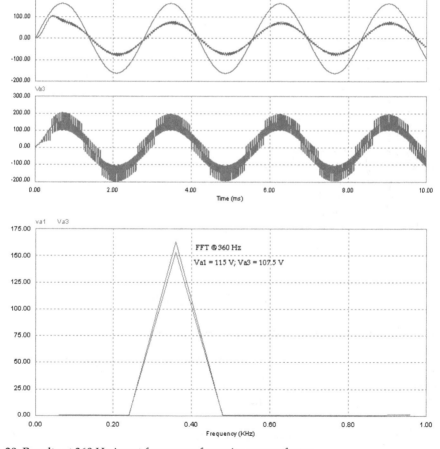

Fig. 20. Results at 360 Hz input frequency for unity power factor.

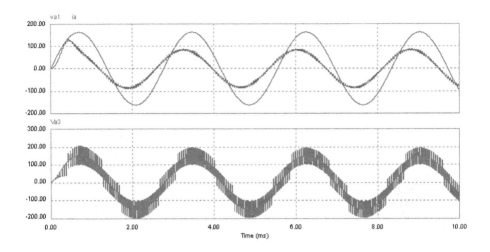

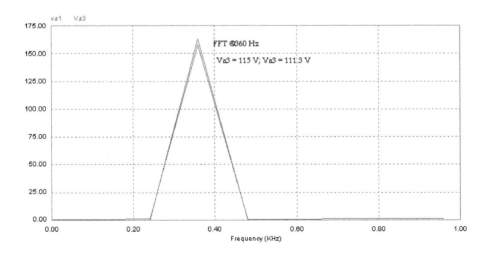

Fig. 21. Results at 360 Hz input frequency for 0.9 power factor.

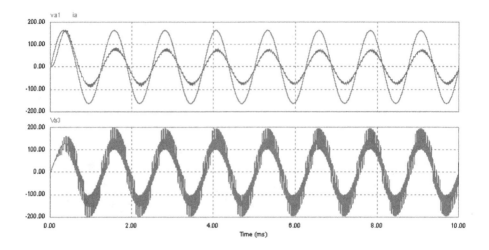

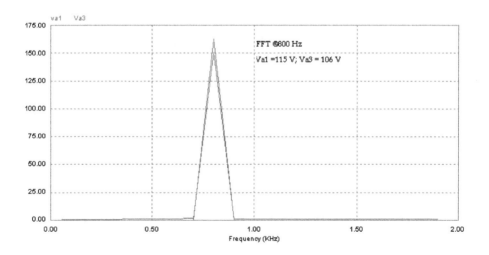

Fig. 22. Results at 800 Hz input frequency for unity power factor.

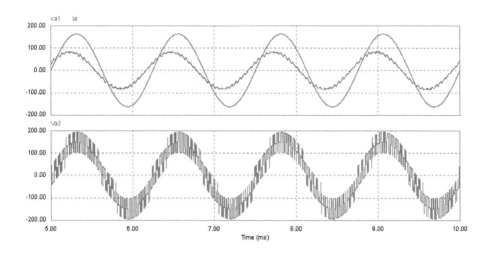

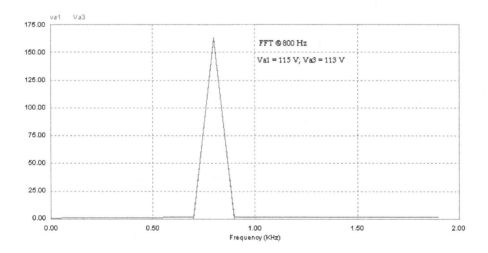

Fig. 23. Results at 800 Hz input frequency for 0.9 power factor.

9. Conclusion

On the basis of the space vector concept a PWM controller was developed. It has been shown that sinusoidal modulation generated in a space vector representation with PI controllers give an adequate performance in steady state and transient condition fast. It has been shown that with the future use of advanced power electronic converters within aircraft equipment, there is the possibility to operate these at variable input frequency and keeping the input current harmonics low.

An AC/DC buck and boost converters with different input frequency and offers low THD (less than 7%) has been described and simulated.

The operation and performance of the proposed topology was verified by simulating a 16 KW with a pure resistive load of 10 Ω and 400 dc voltage for the boost converter and a 3.5 KW with a pure resistive load of 0.5 Ω and 42 dc voltage for the buck converter. The input current is sinusoidal and power factor is unity. The DC voltage is well smoothed.

With the future use of advanced power electronic converters such as active rectifiers and matrix converters within aircraft equipment, there is the possibility to operate these at variable power factor in order to provide system level benefits. These include control of the voltage at the load and improvements in power factor seen at the POR.

10. Acknowledgment

I would like to express my sincere appreciation and respect to the late Prime Minister Rafik El hariri who is entirely responsible for funding my studies in England.

11. References

Faleiro, L. (2005). Beyond the More Electric Aircraft, *Aerospace America*, September 2005, pp 3540

Green, A.; Boys J. & Gates G. (1988). 3-phase voltage sources reversible rectifier, *IEE Proceeding* ,1988,135, pp 362-370, 2002

Green T.; Taha M.; Rahim N.; &Williams B.W. (1997). Three Phase Step-Down Reversible AC-DC Power converter, *IEEE Trans. Power Electron*, 1997,12, pp 319-324

Habetler T. (1993). A space Vector-Bases Rectifier Regulator for AC/DC/AC Converters. *IEEE Trans. Power Electron*, 1993 Vol 8, pp. 30-36

Kazmierkowski M.; Dzeiniakowski M.& Sulkowski W. (1991). Novel space vector based current control for PWM inverters, *IEEE Trans. Power Electron*, 1991, 6, pp 158-166

Taha M (1999). "Power electronics for aircraft application" *Power electronics for demanding applications colloquium, IEE,April 1999, 069 pp 5 -8*

Taha M.; Skinner D.; Gami S.; Holme M. & Raimondi G (2002); *Variable Frequency to constant frequency converter (VFCF) for aircraft application*, PEMD 2002.

Taha M. (2008). Mitigation of Supply Current Distortion in 3- Phase /DC Boost converters For Aircraft Applications, *PEMD 2008.*

Taha, M (2007). Active rectifier using DQ vector control for aircraft power system, *IEMDC 2007* pp 1306-1310

Taha M, Trainer R D (2004); Adaptive reactive power control for aircraft application Power Electronics, Machines and Drives, 2004. (PEMD 2004). Second International Conference on (Conf. Publ. No. 498) 2:469- 474 Vol.2.

Weimer J. (1995). Powe Managemennt and Distribution for More Electric Aircraft, *Proceeding of the 30th Intersociety Energy Conversion Engineering Conference,*pp.273-277

Key Factors in Designing In-Flight Entertainment Systems

Ahmed Akl[1,2,3], Thierry Gayraud[1,2] and Pascal Berthou[1,2]
[1]CNRS-LAAS, Université de Toulouse
[2]UPS, INSA, INP, ISAE; LAAS, F-31077 Toulouse;
[3]College of Engineering, Arab Academy for Science,
Technology, and Maritime Transport, Cairo
[1,2]France
[3]Egypt

1. Introduction

Most of researches concerning *In-Flight Entertainment (IFE)* systems are done on case bases without a global view that encompasses all IFE components. Thus, we try to highlight the key factors of designing IFE system, and showing how its various components can integrate together to provide the required services for all parties involved with the system.

1.1 Background and historical issues

Flight entertainment started before the First World War by the Graf Zeppelin (see Figure 1). This aircraft had a long, thin body with a teardrop shape; it was about 776 feet long and 100 feet in diameter, filled with hydrogen, and the cabin was located under the hull; five engines were fixed to the hull to power the aircraft.

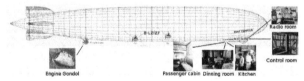

Fig. 1. The Graf Zeppelin aircraft

From the passengers comfort perspective, this model was equipped with a kitchen having electric ovens and a refrigeration unit, a small dinning room, washrooms for men and women, and passenger cabins with a capacity of two passengers each. Unfortunately, the craft was not heated, so passengers were dressing heavy coats and covered with blankets during winter flights. As developments went on, the *"Hindenburg"* aircraft came with heated passenger area, larger dinning room, passengers lounge with a piano as the first audio entertainment, a decorated writing room, a more enhanced passenger cabins, and promenades with seating and windows that can be opened during the flight (Airships.net, Last visit 2011).

In 1949, the *"De Havilland DH 106 Comet"* was the first commercial jet airliner to go into service. It had four jet engines located into the wings. It provided passengers with low-noise pressurized cabin (when compared to propeller-driven airliners), and large windows; hot and cold drinks, and food are serviced through the galley; separate women and men washrooms were available (Davies & Birtles, 1999).

Starting from 1960, *In-Flight Entertainment (IFE)* systems started to attract attention; they were basically a pre-selected audio track that may be accompanied with a film projector. They had shown improvements in both vertical and horizontal dimensions. They expanded horizontally by improving the existing services; audio entertainment moved from using simple audio devices to surround sound and live radio; video display progressed from using a film projector, to CRT displays hanged in the ceiling, to LCD displays dedicated to each passenger. The vertical improvement was noticed through introducing new technologies; cabin telephones allowed passengers to make phone calls during the flight; the system become interactive and allowed passengers to select their own services, while in the past they were forced to follow fixed services; web-based internet services allowed passengers to use some services such as emails and SMS messaging.

The basic idea behind IFE systems was to provide passengers with comfortableness during their long range flights; especially with long transatlantic flights where passengers see nothing but a large blue surface, so that services were initially based on delivering food and drinks to passengers. As passengers demand for more services grows, accompanied with an increase in airlines competition and technology advancement, more services were introduced and modern electronic devices played a remarkable role. This caused a change in the basic concept behind IFE systems; it becomes more than just giving physical comfortableness and providing food. It is extended to provide interactive services that allow passengers to participate as a part of the entertainment process as well as providing business oriented services through connectivity tools. Moreover, it can provide means of health monitoring and physiological comfort.

In recent years, market surveys have revealed a surprising and growing trend in the importance of *IFE* systems with regard to choice of airline. With modern long range aircraft the need for "stop-over" has been reduced, so the duration of flights has also been increased. Air flights, especially long distance, may expose passengers to discomfort and even stress. (Liu, 2007) mentioned that the enclosed environment of the aircraft can cause discomfort or even problems to passengers. This may include psychological and physical discomfort due to cabin pressure, humidity, and continuous engine noise. IFE systems can provide stress reduction entertainment services to the passenger which provides mental distraction to decrease the psychological stress. This can be done by using e-books, video/audio broadcasting, games, internet, and On Demand services. On the other hand, physical problems can range from stiffness and fatigue to the threat of *Deep Vein Thrombosis (DVT)* (Westelaken et al., 2010). IFE systems can provide different solutions such as video guided exercises to decrease fatigue, and seat sensors to monitor the passenger's health status

In fact, passengers from highly heterogeneous pools (i.e., age, gender, ethnicity, etc...) cause an impact on the adaptive interface systems. In non-interactive IFE systems, services (i.e., video and audio contents) are usually implemented based on previous concepts of what passengers may like or require. Using an interactive system based on context-aware services can make passengers more comfortable since they are able to get their own personalized entertainment services. However, such system must be user friendly in terms of easiness to use, and varieties of choice; otherwise, the passenger may get bored and is not able to get the expected satisfaction level.

From the airlines companies' perspective, productivity and profitability are one of the main targets. Achieving these targets is always hindered by the strong competition between companies. Thus, airlines are trying to maximize their attractiveness to get more clients because every empty seat means a revenue loss. IFE systems can play a remarkable role in customer satisfaction and attraction, and it can be used as an efficient portal for in-flight shopping. Moreover, one of the main tasks of aircraft attendants is to keep the passengers calm, unstressed, and to quickly respond to their requests. IFE systems can be a factor of stress elimination, decreasing passenger's movements during the flight, and providing request information quickly to the attendants.

Achieving such level of services requires various technologies and design concepts to be integrated together for implementing such systems. A single networking technology is not capable of providing all types of services. Thus, a good heterogeneous communication network is required to connect different devices and provide multiple services on both system and passenger's levels. For example, a GSM network can provide telephony services; WiFi, Bluetooth, and Infrared to keep passenger's devices connected to the system; LAN and/or *Power Line Communication (PLC)* to form the communication network backbone.

1.2 Chapter structure

Section 2 presents the different types of services provided by IFE systems, and shows the various components which are directly used by passengers as well as the components working at the background, which passengers are not aware of their existence. Section 3 introduces our proposed SysML model that integrates parts of the IFE system to help designers to have a global view of the whole system. Section 4 presents our conclusion. Finally, section 5 discusses future issues of IFE systems.

2. IFE services and components

IFE systems can provide various services for different parties such as airline companies, crew members, and basically passengers. These services are provided through software and hardware components; some components are used directly by passengers, while the others are used indirectly.

2.1 IFE services

IFE services can give solutions for different domains. They can provide health care and monitoring for passengers of health problems, business solutions to advertise products and support business decision making through surveys, and the expected service of entertainment.

2.1.1 Crew services

Although it seems that IFE systems are providing services to passengers only, but it can be extended to provide the cabin attendants with services to facilitate their job. Attendants have to keep a big smile and descent attitude during their work regardless of the current situation, and are burdened with various responsibilities and tasks. We believe that IFE systems can create a dynamic link between passengers and attendants. When an attendant respond to a passenger call, he does not know the reason for the call, so he has to make two moves, one to know the request and the second to fulfill it. An IFE system can allow the passenger to inform the attendant with their request (i.e., drinking water), so that the attendant can finish the service in one move instead of two. Moreover, the IFE system can ask the passenger if he had requested a special meal or not, so the attendant can bring the exact meal to the desired place without moving around with all meals in hand while asking passengers.

The cabin intercommunication service allows the pilot and cabin crew to make announcements to passengers, such as boarding, door closure, take off, turbulence, and landing announcements. These announcements are very important and need to be delivered to all passengers without any interruption; they are usually introduced via a loudspeaker installed in the cabin. If the passenger is wearing his headset, or is not able to understand the announcement language, then few numbers of passengers will comprehend the message. An IFE system can elevate the service through its audio system. When an announcement is introduced while the passenger is running an entertainment service, the entertainment pauses and he hears the announcement through the IFE audio system. Moreover, if it is a standard message such as *"Fasten your seat belt"*, it can be directly translated into the language currently used by the passenger.

Safety demonstrations are used to increase passenger safety awareness. The demonstrations are usually done by crew members. This means that an attendant will stop any current activity and dedicate himself to the demonstration. As an alternative, the IFE system can be used to provide *Aviation safety education for passengers* via multimedia services; insuring accurate instructions, situational awareness, emergency responses, and relevant cabin-safety regulations (Chang & Liao, 2009), so that the attendants can be freed to perform other tasks. Moreover, IFE systems can be used in pre-flight briefing for crew members to improve the quality and availability of information provided to flight crew (Bani-Salameh et al., 2010).

2.1.2 Entertainment services

They are the basic services introduced by IFE systems. They aim at providing multimedia contents for passenger entertainment, audio tracks for different types of music channels, special programs recorded for the airlines, games, and printed media

- *Video on Demand:* As mentioned by (Alamdari, 1999), IFE systems usually include screen-based, audio and communication systems. The screen-based products include video systems enabling passengers to watch movies, news and sports. These systems had progressed into *Video on Demand (VoD)*, allowing passengers to have control when they watch movies. The general VoD problem is to provide a library of movies where multiple clients can view movies according to their own needs in terms of when to start and stop a movie. This can be solved by using an *In-flight Management System* to store the pre-recorded contents on a central server, and streams a specific content to passengers privately.

  The service can be enhanced by using subtitles as a textual version of the running dialogue; it is usually displayed at the bottom of the screen with or without added information to help viewers who are deaf or having hearing difficulties, or people who have accent recognition problems to follow the dialogue. In addition, they can be written in a different language to help people who can not understand the spoken dialogue.

- *Single and multiplayer games:* Video games are another emerging facet of in-flight entertainment. Gaming systems can be networked to allow interactive playing by multiple passengers. Providing high quality gaming in an aircraft cabin environment presents significant engineering challenges. User expectation of video quality and game performance should be considered because many users had experienced sophisticated computer games with multiplayer capabilities, and high quality three dimensional video rendering. Network traffic characteristics associated with computer games should be studied to help in system design; (Kim et al., 2005) measured the traffic of a *Massively Multi-player On-line Role Playing Game (MMORPG)*, showing the differences in traffic between the server and client side. In a *Massively Multiuser Virtual Environment (MMVE)*, where large number of users can interact in real time, consistency management is required to realize a consistent world view for all users. (Itzel et al., 2010) present an approach that identifies users which actually interact with each other in the virtual world, groups them in consistency sessions and synchronizes them at runtime. On the other hand, there is a trend to use wireless networks in IFE systems; the feasibility of using wireless games is studied in different researches (Khan, 2010; Khan et al., 2010; Qi et al., 2009).

- *E-documents:* An in-flight magazine is a free magazine usually placed at the seat back by the airline company. Most airlines are distributing a paper version, and some of them are now distributing their magazines digitally via tablet computer applications. Furthermore, ebooks are widely available electronically with value-added features and search options not available in their print counterparts. Electronic versions are not limited to just text; they may present information in multiple media formats, for example, the text about a type of bird may be accompanied by video depicting the bird in flight and audio featuring its song. Using an electronic version of printed media can change their importance by adding interactive features such as e-commerce services where a passenger can choose his products and buy them instantaneously.

2.1.3 Information services

Air map display provides passengers with up to date information about their travel. They are aware of the plane location and at which part of earth it is passing over. Information telling the outside temperature, speed, altitude, elapsed time, and remaining time gives passengers the sense of movement, because it is difficult at high altitudes, where you can find nothing except blue sky, and sun or moon, to evaluate and sense the aircraft motion. Missing this feeling can be boring for many passengers.

Exterior-view cameras also enable passengers to have the pilot's forward view on take-off and landing on their personal TV screens. The cameras can have different locations. A tail-mounted camera is located in housing atop the vertical stabilizer of the aircraft; it provides a wide-angle view looking forward and typically shows most of the aircraft from above. A belly-mounted camera provides a view looking vertically down, or down at an angle that includes the horizon. A quad-cam belly installation offers a choice of four views covering 360 degrees.

Passengers can pass their time navigating through available entertainment contents to have information about their destination. This can include city maps, sightseeing, languages, and cultural information. Such information will allow passengers to pass a fruitful time and minimize the feeling of being a stranger in a foreign country.

2.1.4 E-business services

Airborne internet communications allows passengers and crew members to use their own WiFi enabled devices, such as laptops, smart phones and PDAs, to surf the Web, send and receive in-flight e-mail with attachments, Instant Message, and access their corporate VPN. Many companies are offering solutions to provide passengers with Internet connectivity. FlyNet (FlyNet, Last visit 2011) is an example for onboard communication service provided by Lufthansa to allow passengers to connect to the Internet during their flight. ROW44 (ROW44, 2011) provides a satellite-based connectivity system that allows airlines to offer uninterrupted broadband service

Mobile phones are one of the most demanded devices by passengers. Many passengers, especially businessmen, are welling to make calls through their personal mobile phone during their flight. However, there are doubts that cell phone signals may endanger aircraft safety by interfering with navigational systems. To overcome this situation, different techniques (i.e., (AeroMobile, Last visit 2011)) were introduced to the market, where an on-board pico cell can connect the mobile phones to the ground stations through the satellite link and managing the signal strength to insure that there is no interference with the navigational systems.

On-board conferencing can turn wasted flight time into productive time for traveling teams of salespersons. Also, it will reduce the effort done by passengers to trade seats after boarding to bring their group together. With the addition of a headset with *Active Noise Cancellation*, the experience can be extended to conversing with someone in the next seat, due to the reduction of ambient noise.

Personal Electronic Devices (PEDs) such as laptop computers (including WiFi and Bluetooth enabled devices), PDAs (without mobile phones), personal music (i.e., iPods), iPads, ebooks and electronic game devices are electronic devices that can be used when the aircraft seat belt sign is extinguished after take-off and turned off during landing. On the other hand, other PEDs using radio transmission such as walkie-talkies, two-way pagers, or global positioning systems are prohibited at all stages of flight, as it may interfere with the aircraft communication and navigation systems.

Power outlets are hardly reached by passenger during traveling to their destination. Spending too much time without a power source can cause PEDs to run out of power, and causing passengers to be frustrated. As a solution for such situation, airlines (AmericanAirlines, 2011; Qantas, 2011) add power outlets to passenger seats. These outlets are usually present in first and business class seats. For safety reasons, some outlets are designed to provide 110 Volt (60 Hz) with 75 watts, however, this may be unsuitable for PCs that consumes more power. Other companies provide 15 volt cigarette lighter outlet, which needs an adapter to connect devices.

2.1.5 E-commerce services

In-flight shopping is dragging more attention from airlines as it is considered as a source of revenue, and a way for passengers to utilize their flight time. (Liou, 2011) presented passenger attitude towards in-flight shopping. He mentioned that customer's convenience increases when the shopping process takes less time, less effort in planning ahead, and less physical effort to obtain the product or service. Moreover, many factors can affect the decision making process (i.e., to buy a product); this includes pre-purchase information searching, and evaluation of alternatives.

An IFE system can be a remarkable factor for in-flight shopping. It can increase passenger convenience and facilitate decision making. An electronic catalogue viewed through the IFE display unit can provide search options that allow passengers to find other alternatives and make his own comparisons, and it can provide him with exhaustive information about the product. In turn, this will allow passenger to plan ahead without making two much physical effort, and in a relatively shorter time than making the same process in a paper document or through discussion with a crew member. Furthermore, the IFE system can play an extra ordinary role to e-commerce, not only for in-flight shopping, but also for shopping outside the flight. The IFE system can be connected to ground commercial services, so that the passenger can buy products or services (i.e., transport tickets, and duty free products), and receive them directly when he reaches his destination. In addition, multimedia advertising can attract companies to use it as a way to reach passengers.

Surveying is an important part of market evaluation. (Balcombe et al., 2009) held a survey to determine passenger's *Willingness To Pay (WTP)* for in-flight service and comfort level. The survey focused on seat comfort, meal provision, bar service, ticket price, entertainment (i.e., overhead screens for pre-set programs). He reported that older passengers are WTP more for seat comfort, while younger passengers are WTP more for bar and screen services.

However, performing such surveys is very tedious and difficult. (Aksoy et al., 2003) held a survey to evaluate Airline service marketing by domestic and foreign firms from customers viewpoint. The usable responses were 1014 out of 1350 responses, producing a 75.1% response rate. An IFE system can be an effective tool to increase the response rate, where an electronic version of the survey can guarantee that more passengers will participate, erroneous answers can be reduced, and analyzing the results becomes faster and accurate.

2.1.6 Health services

An elevated type of services, which IFE can provide, is health services. Flight conditions may cause the cabin environment to be tough; especially for persons who can face ill conditions. Flight duration, dehydration, pressure, engine noise, and other factors can be reasons of physical and/or psychological problems. A sensory system integrated in IFE system can provide a way to sense bad health conditions of passengers having health problem, and either inform the crew members or perform an action to reduce the effect.

(Schumm et al., 2010) and (Westelaken et al., 2010) suggest solutions based on sensory systems embedded in passenger's seat to sense his current status. (Schumm et al., 2010) introduce the design of smart seat containing sensors to measure *Electrocardiogram (ECG)*, *Electrodermal Activity (EDA)*, respiration, and skin temperature. These measured values can give a good indication about physical and psychological state. ECG is measured in two ways; without skin contact through sensors embedded in the backrest, and with a sensor fixed on the index finger. The second type is more obtrusive, but is more reliable. The same fixation system to the finger includes the EDA, and temperature sensors. The passenger movements can affect the reading quality, so a 3-axis accelerometer is added to compensate the errors. Respiration level is detected through sensors fixed in the seatbelt. The combined reading of these sensors can give a good indication about the passenger's health status.

Physical exercises can reduce physical stress and fatigue. However, the challenge is how to stimulate passengers to do them. (Westelaken et al., 2010) introduces a solution to reduce physical and psychological stress by detecting body movements and gestures to be used as an input for interactive applications in the IFE system. The basic idea for implementing these applications is to allow the passenger to participate in a gaming activity. His movements are captured as inputs for the chosen game. Three techniques were introduced to capture movements; sensors integrated in the floor, sensors integrated in the seat, and video-based gesture recognition. However, each of these techniques has its own pros and cons which need more investigation.

For passengers of special health needs, IFE system can be an effective tool to relief their pain. Passengers of *Spinal Cord Injury (SCI)* are not able to sense pressure acting on certain parts of their body that are cut off nervous system. This may increase the risk of decubitus ulcer, especially for long flights, where passengers may sit for several hours. (Tan, Chen, Verbunt, Bartneck & Rauterberg, 2009) proposed an *Adaptive Posture Advisory System (APAS)* for people of SCI. The passenger's seat plays a great role by having various sensors and actuators. Sensors are used as input source for a central processor connected to a database which is used to record passenger's sitting behavior and conditions. The suitable decision is taken and sent to the actuator to change the seat shape, and softness. This system helps SCI passengers to reposition their sitting posture to shift the points under pressure so that decubitus ulcer risk is minimized.

2.2 IFE components

The IFE components can be categorized into passenger and system components. In (Akl et al., 2011), we identified passenger components as the devices that the passenger uses directly to achieve a service, and system components as the components which are provided by the system and used indirectly by the passenger.

2.2.1 Passenger components

Passenger components are usually designed to be very simple and familiar in appearance and functionality in order to allow passengers of different background to use them; such as display units, remote controls, seat control buttons, headphones, etc...

2.2.1.1 Passenger seat

From the first sight, the passenger's seat may seem to be out of the scope of IFE systems, which are basically designed for entertainment. However, a deep look shows the contrary since passenger's seat is one of the main comfortableness components; especially when we consider that it is the place where the passenger spends most of his travel time. From one side, a poorly designed seat can causes discomfort, which can be extended to a musculoskeletal disorders regardless of the presence of any entertainment or stress reduction techniques; imagine the stay on such a seat for three or two hours, you will think in nothing except the time when the flight ends. Furthermore, when the passenger sits upright and inactive for a long period of time, he may be exposed to several health hazards. The central blood vessels in his legs can be compressed, making it harder for the blood to get back to his heart. Muscles can become tense, resulting in backaches and a feeling of excessive fatigue during, and even after the flight. The normal body mechanism for returning fluid to the heart can be inhibited and gravity can cause the fluid to collect in the feet, resulting in swollen feet after a long flight.

From another side, modern technologies can be used to elevate seat entertainment and comfortable role. Thus, we propose two terms, *Passive Seat (PS)*, and *Active Seat (AS)*. The *PS* is providing the service through its own structural design without any interaction with the passenger. The *AS* is providing the service in response to an intentional or unintentional input captured from the passenger.

- **Passive Seat:** (Nadadur & Parkinson, 2009) discussed different seat design problems. Airlines are aiming at increasing the seats density inside the cabin to increase their revenue. However, such approach diminishes the comfortableness factors in seat design. Increasing the seats density negatively affects the seat pitch, causing a decrease in the passenger's leg room (see Figure 2), which is considered as an important factor especially for tall passengers. He also mentioned that passengers should minimize the pressure between their lower thighs and the surface of the seat to prevent the occurrence of *Deep Vein Thrombosis (DVT)*. This can be achieved by keeping the knees height greater than the seat's height. A design contradiction here arises because lower seat height requires more leg room causing seat pitch to increase, and consequently seats density will decrease. On the other hand, increasing the seat height increases discomfort and the probability to have DVT problems. To find a compromise between these contradictions he proposed a mathematical solution to embed the passenger comfort as a design parameter and link it with the passenger's willingness to pay higher prices. (Vink, 2011) introduced other factors to be considered during seat design such as wider seats, adjustable headrests, space under the armrest, backrest angle, and ideal distribution of pressure over body parts. A better pressure distribution can be achieved by using support under the front part of the legs to spread the load, and ergonomic design of seat back and seat pan. Also, a well designed headrest and neck rest can increase the comfort feeling.

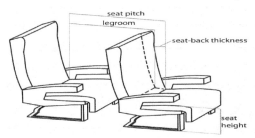

Fig. 2. Some aircraft seat design parameters from (Nadadur & Parkinson, 2009)

Sleeping and sitting posture is an important factor for passenger's comfort especially in long haul flights and it can affect the pressure distribution over the body. (Tan, Iaeng, Chen, Kimman & Rauterberg, 2009) held an analysis of passengers' postures in the economy class to help in seat design. In their study, they identified seven different sleeping positions for passengers. When considering the anthropometry differences between humans of different origins, we can say that it is difficult for a passive seat to achieve all comfort positions of different postures for all passengers, so an active seat with adjustable moving parts is usually required.

- *Active Seat*: A *Passive Seat* provides services of static features. On the contrary, an *Active Seat* is able to get an input from the passenger to change the service it provides. The input can be an activity to change the angle or position of adjustable parts of the seat; for example, the passenger can freely set the backrest angle or adjust the height of headrest to match his posture. In business class, the seat can accommodate a variety of postures for different activities such as watching TV, reading, sleeping, etc... Figure 3 shows a simple mechanical button (in economic class) for changing backrest angle Vs an electronic buttons (in business class) that can easily change the orientation of different parts of the seat using embedded motors.

Fig. 3. Electronic Vs Mechanical seat adjustment

In economy class, the degrees of freedom of an *Active Seat* are very limited where minimal parts are allowed to change their orientation due to limited space. For example, the armrest can be moved from the horizontal position to the vertical position to give more space, and the backrest angle can be changed to increase the body inclination and reduce the pressure exerted on the back. However, the inclination angle is usually very small in order not to reduce leg space of the behind seat. On the contrary, the business class seat is featured by large spaces; thus, different parts can be reoriented easily. A premium seat may be in a pod and capable of opening out into a flat sleeping configuration or folding up into a seat for take-off and landing. Moreover, it includes more amenities such as power, task lighting, and has also a design trend towards a higher level of privacy.

2.2.1.2 Visual display units

A *Visual Display Unit (VDU)* is the principal component in the entertainment process. It is the main interface between passengers and the IFE system, as well as their ability to provide interactive services. There are different types of VDUs. At the very beginning, *Cathode Ray Tube (CRT)* displays were used. Although they were able to provide the required service at that time, but were suffering of many drawbacks. They were relatively large in size and heavy in weight, so they were used as a shared display between a set of seats. Furthermore, the ambient lighting may affect the clearness of images. As technology advances, *Liquid Crystal Display (LCD)* units were introduced. They are small in size and light in weight. These characteristics helped greatly in introducing *Video on Demand (VoD)* service, where each passenger has his own display unit to watch his selected items. At the same time LCDs can still be used as shared displays. Nowadays, displays are equipped with an extra feature that allowed them to be used as input devices. Touch screens allow users to choose their own selections by touching the screen in the appropriate location.

Although a normal VDU is usually sufficient to display the required contents, certain services may have special needs. Table 1 shows the characteristics required to display different media services. With respect to the display quality, Video games do not need high resolution for their images since small moving objects are the main constitute of Video games. On the contrary, movies and virtual reality applications need high resolution to present their high quality images. The interactive feature of Video games and Virtual Reality applications require special input devices, since touch screens are usually suitable for simple selections and not for quick repetitive pressing.

Service	Realistic	Interactive	Immersive	Detailed Character
Video Games	No	Yes	No	Yes
Movies	Yes	No	No	Yes
Virtual reality	Yes	Yes	Yes	Yes

Table 1. Various Display requirements

The VDU location depends on the philosophy of the installed IFE system. If the system is going to present the pre-selected media without any intervention from the user, then a global VDU is installed in the cabin ceiling (see Figure 4(d)). If VoD service are presented with user interaction to select his own media contents, then each passenger seat is provided with a private VDU fixed in the back of the front seat (see Figure 4(a)). Furthermore, seats of special locations such as seats of first row or in the business class may have special VDU placement (see Figure 4(b) & 4(c)).

The VDU viewing angle is an important satisfaction factor. The viewing angle of VDUs fixed at the back of the front seat may change when the front passenger changes the position of his seat back, so that VDUs are usually fixed on a pivot to allow the user to change their inclination; otherwise, the user has to move his head to a fixed position to be able to view the VDU. Another solution is to fix the VDU on a movable axis to give the VDU different degrees of freedom (see Figure 4(b))

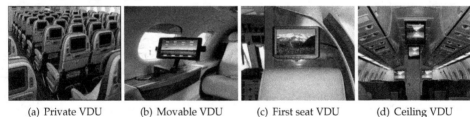

(a) Private VDU   (b) Movable VDU   (c) First seat VDU   (d) Ceiling VDU

Fig. 4. Different VDU placements

2.2.1.3 Remote control

As IFE systems are becoming more and more interactive, a *Remote Control Device (RCD)* is needed to control the surrounding devices. It should be compact and easily held. Moreover, the pocket holding the RCD has to be placed in a way that makes it easily reached and not to affect passenger comfort. At the beginning, RCD used to be fixed aside to the VDU at the back of the front seat. This orientation introduced a problem when the passenger setting beside the window wants to move to the corridor; where all his neighbors have to replace their RCDs to allow him to pass. To overcome this problem, RCDs are now connected to their VDUs through wires passing via their seat. Using wireless technology can minimize such physical complexity (Akl et al., 2011).

Furthermore, passengers of no knowledge about using modern technology must be able to use RDCs easily. Usual control buttons (i.e., Volume, Rewind, Forward, etc...) are known for almost everyone; especial purpose controls such as *Settings*, and *Mode* can be carefully manipulated and, if used, to be provided by explanatory information when possible.

2.2.1.4 Noise canceling headphones

Headphones are used to privatize audio contents, so that each passenger can listen to his own selection without annoying his neighbors or being affected by the surrounding noise. Ordinary headphones are usually enough to do the job. However, modern technology can elevate the service level, by introducing active headphones capable of reducing the effect of surrounding noise (see Figure 5).

Generally, headphone ear cups have passive absorption capability which allows them to block some high frequency noise. However, they are not efficient for attenuating low frequency noise. A *Noise Canceling Headphones (NCH)* can reduce the noise through active noise cancellation techniques (Chang & Li, 2011)

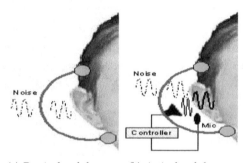

(a) Passiveheadphones   (b) Activeheadphones

Fig. 5. Headphones

2.2.1.5 Personal Electronic Device (PED)

Nowadays, people are getting more sticky to their *Personal Electronic Devices (PEDs)* such as laptop, mobile phone, and PDA, so most passengers are traveling with their PEDs. Connecting PEDs do not require special interfaces since modern IFE systems are moving towards wireless communication such as WiFi, Bluetooth, and IrDA, which are already used in most PEDs.

Using PEDs can have several advantages for both Airlines and passengers. Passengers will be able to use their devices to interact with the IFE system. They do not need to use or investigate unknown devices. Also, they can utilize their own data if the system permits them. Furthermore, if the IFE contents can be copied, the passenger can continue it at his hotel.

From the airlines perspective, PEDs can be used to save some dedicated devices of IFE systems. It is cheaper for airlines to remove expensive seatback monitors and let passengers to use their own devices; this is a good option for airlines offering cheap flights. Many companies (Lufthansa, 2011; Thales, 2011) are now offering broadband communication for PEDs.

2.2.2 System components

System components are usually complex to be able to handle the services while keeping simplicity of passenger components. Furthermore, the cabin environment is strict in terms of safety and imposed constrains. These characteristics encouraged the solution of using multiple technologies to form a heterogeneous system where each technology provides a solution for a part of the problem.

A context-aware IFE system can increase passenger satisfaction level. If there are many choices and the interaction design is poor, the passenger tends to get disoriented and is not able to achieve the most appealing contents. This is because most IFE systems are user adaptive systems where the user initiates system adaptation to get his personalized contents. (Liu & Rauterberg, 2007) showed the main architectural components to make a context-aware IFE system which can provide the passenger with entertainment contents based on his personal demographic information, activity, physical and psychological states if the passenger was in stress. Furthermore, the passenger is able to decline the proposed contents, and create his personalized contents.

For IFE networking, wireless technology can introduce different solutions to solve many existing problems as well as providing new services. Nowadays, wired networks are the principal technology of implementing IFE systems. Ethernet is currently the standard for wired communication in different fields. (Thompson, 2004) showed that it is characterized by interesting features such as good communication performance, scalability, high availability, and resistivity to external noise. Using off shelf technologies such as routers can reduce the costs of networking inside the cabin. In spite of all these advantages, IFE system designers are welling to exchange it -or part of it- by wireless technology to achieve more targets. Ethernet cabling is considered a burden for aircraft design because lighter aircrafts consume less fuel, and it imposes difficulties on easiness of reconfiguration and maintenance of the cabin (Akl et al., 2011). Accordingly, using different technologies within the same communication network can introduce a solution to the limitations of using each of them individually.

2.2.2.1 WiFi and Bluetooth

WiFi is a well known technology used in different commercial, industrial, and home devices. It can easily coexist with other technologies to form a heterogeneous network (Niebla, 2003). Moreover, (Lansford et al., 2001) stated that WiFi and Bluetooth technologies are two complementary not a competing technologies. They can cooperate together to provide users with different connecting services.

However, using large number of wireless devices in a very narrow metallic tunnel like the cabin has a dramatic effect on network performance. Furthermore, a major concern for using wireless devices in aircraft cabin is their interference with the aircraft communication and navigation system, especially unintended interference from passenger's *Personal Electronic Devices(PED)*. (Holzbock et al., 2004) said that the installed navigation and communication systems on the aircraft are designed to be sensitive to electromagnetic signals, so they can be protected against passenger's emitters by means of frequency separation. In addition, (Jahn & Holzbock, 2003) mentioned that there are two types of PEDs interference, intentional and spurious. The former is the emissions used to transmit data over the PED allocated frequency band. The latter is the emissions due to the RF noise level. However, indoor channel models mainly investigate office or home environments, thus these models may not be appropriate for modeling an aircraft cabin channel. Attenuation of walls and multi path effects in a normal indoor environment are effects, which are not expected to be comparable to the effect of the higher obstacle density in a metallic tunnel. The elongated structure of a cabin causes smaller losses, than that expected in other type of room shapes. However, the power addition of local signal paths can lead to fading of the signal in particular points. In addition, small movements of the receiver can have a substantial effect on reception. The same opinion was emphasized by (Diaz & Esquitino, 2004).

Different efforts were held to overcome this problem, (Youssef et al., 2004) used the commercial software package *Wireless Insite* to model the electromagnetic propagation of different wireless access points inside different types of aircrafts. (Moraitis et al., 2009) held a measurement campaign inside a Boeing 737-400 aircraft to obtain a propagation development model for three different frequencies, 1.8, 2.1, and 2.45GHz which represent the GSM, UMTS, and WLAN and Bluetooth technologies, respectively. Nowadays, many airline companies allows WiFi devices on their aircrafts such as Lufthanza (FlyNet, Last visit 2011), and Delta Airlines (DeltaAirline, Last visit 2011).

2.2.2.2 Wireless Universal Serial Bus (WUSB)

Universal Serial Bus (USB) technology allows different peripherals to be connected to the same PC more easily and efficiently than other technologies such as serial and parallel ports. However, cables are still needed to connect the devices. This raised the issue of *Wireless USB (WUSB)* where devices can have the same connectivity through a wireless technology. (Leavitt, 2007) stated that although it is difficult to achieve a wireless performance similar to wired USB, but the rapid improvements in radio communication can make WUSB a competent rival. It is based on the *Ultra Wide Band (UWB)* technology. In Europe, it supports a frequency range from 3.1 to 4.8 GHz. Moreover, (Udar et al., 2007) mentioned that UWB communication is suitable for short range communications, which can be extended by the use of mesh networks. Although WUSB was designed to satisfy client needs, but it can also be used in a data centre environment. They discussed how WUSB characteristics can match such environment. This application can be of a great help in IFE systems, which strive to massive data communication to support multimedia services and minimizing connection cables. Moreover, (Sohn et al., 2008) discussed the design issues related to WUSB. He stated that WUSB can support up to 480 Mbps, but in real world it does not give the promised values; and they showed the effect of design parameters on device performance.

2.2.2.3 PowerLine Communication (PLC)

A PLC network can be used to convey data signals over cables dedicated to carry electrical power; where PLC modems are used to convert data from the digital signal level to the high power level; and vice versa. Using an existing wiring infrastructure can dramatically reduce costs and effort for setting up a communication network. Moreover, it can decrease the time needed for reconfiguring cabin layout since less cables are going to be relocated. However, such technology suffers from different problems. A power line cable works as an antenna that can produce *Electromagnetic Emissions (EME)*. Thus, a PLC device must be *Electromagnetic Compatible (EMC)* to the surrounding environment. This means that it must not produce intolerable EME, and not to be susceptible to them. To overcome this problem, the transmission power should not be high in order not to disturb other communicating devices (Hrasnica et al., 2004). However, working on a limited power signal makes the system sensitive for external noise. In spite of this, the PLC devices can work without concerns of external interference due to two reasons. Firstly, the PLC network is divided into segments; this minimizes signal attenuation. Secondly, all cabin devices are designed according to strict rules that prevent EME high enough to interfere with the surrounding devices. (Akl et al., 2010) presented a PLC network dedicated for IFE systems to replace part of the wired communication network, where two PLC devices were used; *Power Line Head Box (PLHB) and Power Line Box (PLB)*. PLHB connects the two terminals of the power line to connect data servers with seats. Each PLHB service a group of seats, which are equipped with PLB per seat (see Figure 6).

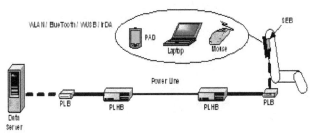

Fig. 6. Heterogeneous network architecture

2.2.2.4 GSM

For several years the aircraft industry has been looking for a technology to provide, at a reasonable cost, an onboard phone service (see Figure 7). Nevertheless, some technical hitches make successful calls via the terrestrial *Global System for Mobile Communications (GSM)* network impossible. The mobiles are unable to make reliable contact with ground-based base stations, so they would transmit with maximum RF power and these RF fields could potentially cause interference with the aircraft communications systems. On the other hand, the high speed of the aircraft causes frequent handover from cell to cell, and in extreme cases could even cause degradation of terrestrial services due to the large amount of control signaling required in managing these handovers. In order to avoid these problems and allow airline passengers to use their own mobile terminals during certain stages of flight, a novel approach called *GSM On-Board (GSMOB)* is used. The GSMOB system consists of a low power base station carried on board the aircraft itself, and an associated unit emitting radio noise in the GSM band, raising the noise floor above the signal level originated by ground base stations. Thus mobiles activated at cruising altitude do not see any terrestrial network signal, but only the aircraft-originated cell. This way, the power level needed is low, which reduces the interference with aircraft systems.

The AeroMobile (AeroMobile, Last visit 2011) is a GSM service provider for the aviation industry that allows passengers to use their mobile phones and devices safely during the flight. Passengers can connect to an AeroMobile pico cell located inside the craft which relays text messages and calls to a satellite link which sends them to the ground network. The AeroMobile system manages all the cellular devices onboard. This system is adopted by Panasonic to be part of its in-flight cellular phone component.

2.2.2.5 Satellite communication

In-cabin communication can be extended by being connected to terrestrial networks through satellite links (see Figure 7). Using satellite channels allow passengers to use their mobile phones, send emails, access internet, and achieve online entertainment services. However, the satellite link is considered as the connection bottleneck, so traffic flow in and out of the cabin must be analyzed (Niebla, 2003). (Radzik et al., 2008) performed a satellite system performance assessment for IFE system and *Air Traffic Control (ATC)*, where the satellite link can be shared between IFE and ATC streams. (Holzbock et al., 2004) presented, in details, two systems that allow in-cabin communication to be connected to assessment networks; the ABATE system (1996-1998), and the WirelessCabin system (2002-2004). Another recent project is the E-CAB project (ECAB, Last visit 2011) which was held by Airbus.

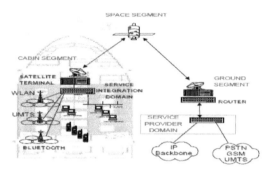

Fig. 7. Satellite link from (Niebla, 2003)

3. Design and evaluation of modern IFE systems

To design an IFE system, different types of requirements need to be defined and constrains must be considered. It is not just adding some entertainment devices, but it is a system which will be located in a very strict environment. This system will have an impact on passengers, airlines, and aircraft design. Therefore, a formal modeling of IFE systems is a paramount need which can be achieved through *System Modeling Language (SysML)*. SysML is a modeling language for representing systems and product architectures, as well as their behavior and functionalities. It is an important tool to have an understanding of a system to prevent complex failure modes leading to costly product recalls. Furthermore, it uses generic language, which is not specific to any engineering discipline, able to present the incremental details of system modeling. Modeling starts by gathering the required functionality until reaching the complex system model. This is achieved by presenting its sub-system structures, and showing their behavior of interacting together as well as with external system. However, we have to stress on the fact that there is no optimum model for any system, but we can have a good model. A good model is the one that fulfills all of system functional and non-functional requirements.

3.1 Proposed IFE model

In this section, we propose a SysML model that takes us through a step by step design process as a systematic design approach to help designers to handle such complex system. The model will show system components, the involved actors, and their interactions with the system. We believe that the model can give the designer an idea on how to adapt his own IFE system to achieve the expected services.

A real IFE system is a large system, where its model can not be fully presented in a book chapter, so we will consider a small case study, and stress on the basic steps and techniques that should be considered during the design process.

Our case study is based on the work done by (Loureiro & Anzaloni, 2011), and our previous work in (Akl et al., 2010) to model the part related to the VoD service and the PLC network. (Loureiro & Anzaloni, 2011) introduced a peer-to-peer networking approach for using VoD for IFE systems and propose two solutions for the problem of content searching in such network. We chose their work because it is a recent research that presents two different techniques to distribute video content over a peer-to-peer network rather than using traditional client-server architecture. The peer-to-peer approach allows passenger IFE units to monitor, store, and serve media contents to each other. This can be achieved by having a *Distribution Table (DT)* containing the video file information (i.e., file ID and IP of storing peer). The work is based on how to build and update the DT. In (Akl et al., 2010), we proposed using a PLC network to replace traditional LAN (see Figure 6). The PLC system consists of a *Power Line Head Box (PLHB)* and a *Power Line Box (PLB)*, where the PLHB connects the two terminals of the power line. Each PLHB service a group of seats which are equipped with PLB per seat. The PLB is responsible for distributing the signal received by the PLHB to the seat attached devices. Each PLHB can support up to 20 PLBs at a rate of 3480 bit/sec each. We will use the model to verify if the technique proposed by (Loureiro & Anzaloni, 2011) can be supported by the PLC network or not.

3.1.1 Use Case diagram

A *Use Case* describes system functionality in terms of how its users (i.e., passengers, crew) use the system to achieve the needed targets. It represents a high level of abstraction to model IFE requirements and interaction with users. Consequently, it typically covers scenarios through which stake holders (i.e., actors) can use the IFE system. Hull et al. (2011) stated that "*A stake holder is an individual, group of people, organization or other entity that has a direct or indirect interest (or stake) in a system*".

In a Use Case, the system boundaries are identified by a square box to decide what belongs to the system and what does not. For example, a GPS device that provides the IFE system with data used in a map display is considered as a part of an external system (i.e., navigational system).

Figure 8 presents a *Use Case diagram* for our proposed IFE model. There are seven actors; passengers, crew members, a navigational system, a cabin environment, maintenance personnel, airline company, and avionic regulations. The IFE system is enclosed inside the box representing the system boundary. The oval shapes show the interactions of each actor with the system. These interactions are related together through different relations (i.e.,

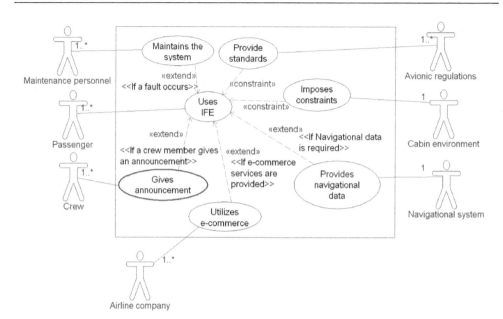

Fig. 8. IFE system Use Case diagram

extended, constrain). *Extend* relationship identify an *extending use case* which is a fragment of functionality that is not considered part of the normal base use case functionality. *Constraint* relationship shows constraints imposed on the system.

The base Use Case is *"Uses IFE system"* which is directly utilized by passengers. It represents the utilization of IFE components (see section 2.2). Its functionality can be extended when a crew member gives an announcement (see section 2.1.1), or the navigational system provides data, or a maintenance personnel performs a maintenance action. Constraints comprise the difficulties imposed by cabin environment, and the standards provided by avionic regulations (i.e., ARINC standard 808, RTCA DO-160E). The next step is to model the requirements needed by stakeholders.

3.1.2 Requirements model

We present a part of the basic requirements related to the entertainment service that can exist in any IFE systems. These requirements are categorized as functional and non-functional requirements. This step helps designers to highlight the basic features of their system.

Defining system requirements seems easy, but in fact, it is not. The defining requirements process is divided into several steps. Firstly, to define stake holders. Second, to start a requirement gathering process, where requirements are collected from stakeholders. Finally, requirements are organized according to well defined rules that guarantee certain requirement characteristics which are essential for requirement analysis. For more information about requirement engineering, we refer readers to (Hull et al., 2011; Young, 2004).

We will assume that the first and second steps are already done, so that our IFE system requirements are already gathered from stakeholders, and we will classify them as functional and non-functional requirements.

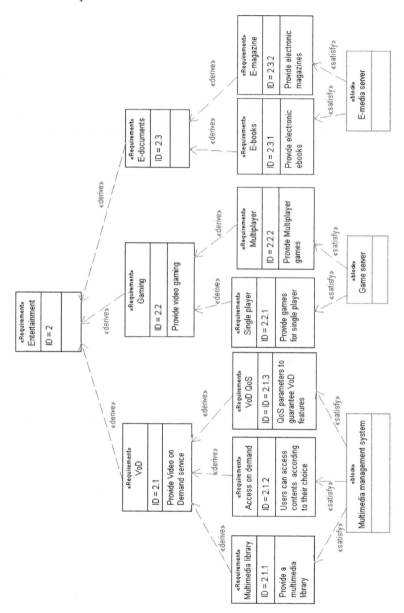

Fig. 9. Requirement diagram of entertainment specifications

3.1.2.1 Functional requirements

Functional requirements describe what the system is supposed to do by defining its behavior (i.e., functions and services). For an IFE system, this includes the different services provided to passengers, and airlines companies (see section 2.1). For each service, there is a dedicated requirement diagram. A group of related requirements are called a specification.

Figure 9 presents the specifications of entertainment service. Each block represents a requirement; showing its name, ID number, and text explaining the purpose of the requirement. The *Derive* relationship shows sub-requirements needed to fulfill the parent requirement. For example, our entertainment service will include VoD, Gaming, and E-documents services. The VoD service will be fulfilled through a *Multimedia Library* to store the VoD contents, and an *Access on Demand* capability. A system component is responsible for satisfying (i.e., represented by the *Satisfy* relationship) these requirements; it is named the *Multimedia Management System*. If necessary, the last level of requirements can decompose into finer levels of derived requirements to show more details of the system. The *Distribution Table* technique (Loureiro & Anzaloni, 2011) will be used to satisfy part of the requirements of VoD service

3.1.2.2 Non-Functional requirements

Non-functional requirements describe constraints and qualities. *Qualities* are properties or characteristics of the system that will affect user's degree of satisfaction. This includes maintainability, reliability, security, and safety issues. Designers usually focus on system functionality and may lately consider the non-functional requirements during the design process. Failing to achieve non-functional requirements may lead to a functional system with undesirable level of satisfaction.

Figure 9 shows QoS parameters as the non-functional requirements needed for the VoD service. (Loureiro & Anzaloni, 2011) identified two main parameters ρ and θ to define the required transmission. ρ and θ represent the amount of information (bytes) that needs to be transmitted across the application layer of the network during system startup, and system normal operation, respectively. They are presented in our model as constraints (explained further in section 3.1.3). ρ and θ are defined as:

$$\rho = nF(c_6 + c_7L) \tag{1}$$

$$\theta = c_5n \tag{2}$$

where n is the total number of peers, F is the number of messages sent between two nodes. L is the number of video files stored in the node's local storage, and c_5, c_6, and c_7 are constants. The next step is to model the system components that satisfy these requirements.

3.1.3 Structural model

Block Definition Diagram realizes the structural aspects of the model. It shows which components exist in the system, and the relation between them. It is formalized and reconciled with both behavior model and requirements. Blocks are used to present components; they are connected through relations, and ports to describe the points at which a block interacts with another block.

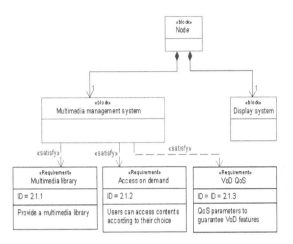

Fig. 10. Block diagram of node structure and satisfied requirements

There are two types of ports: *Flow ports* specify what can flow in and out of blocks (i.e., data or physical items), and *Standard ports* that specify the types of services that a block either require or provide.

Figure 10 shows the main blocks of each node and its relation with the requirements depicted in figure 9. It consists of two main blocks; *Multimedia Management System* and *Display System*. The former manages the multimedia contents, while the later is responsible for displaying multimedia contents and receiving passenger selections. The figure does not show the requirements satisfied by the *Display system* block because we are only interested in the requirements of entertainment service shown in figure 9.

Figure 11 shows the node composition, and its relation with other components (i.e., *Networking System* block). Operations are listed in the *Operations* compartment of the block. However, for readability reasons, we only show the operations of *Multimedia management system* block. *Networking system* is responsible for handling communication between nodes. This is done through the PLHB component that connects different groups of PLBs. The *Multimedia Management System* block consists of three managers; *Content Search Manager, Local Storage Manager*, and *Content Selection Manager*. The *Local Storage Manager* handles the local multimedia contents, and defines its location inside the storage device. The *Content Selection Manager* receives the selection request from the *Display System* block, and send back the media content after being received from the *Content Search Manager*. The *Content Search Manager* searches for the requested item in the way mentioned by (Loureiro & Anzaloni, 2011) (the behavior of this technique is modeled in the next section). If the content is not stored locally, a search will be retrieved from neighboring nodes by communicating through the PLB component.

Parametric diagram uses constraint blocks that allow to define and use various system constraints. These constraints represent rules that can constraint system properties, or define rules that the system must conform to. A constraint block consists of constraint name and constraint formula. All variables or constants defined in the formula are linked to the block through an *Attribute* box or through an input from another block.

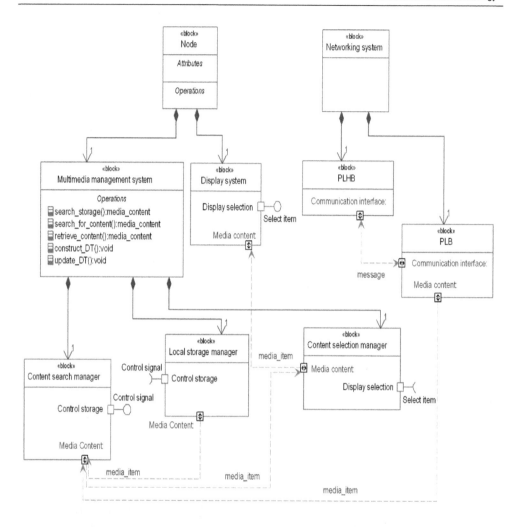

Fig. 11. Block diagram of control signals and flow items

Figure 12 shows three main constraint blocks; *Peer-to-Peer Transmission Rate, PLC Parameters,* and *Acceptance Criteria*. The *Peer-to-Peer Transmission rate* defines three formulas (as mentioned in (Loureiro & Anzaloni, 2011)). The PLC parameters define the PLHB maximum bandwidth, as mentioned in (Akl et al., 2010). The output of the two constraints are used to determine the validity of *Criteria 1. Criteria 1* is valid when the PLHB maximum bandwidth is greater than the rate of data transmission *B*. This means that the PLC network is able to handle the traffic generated to update the distribution table. *Criteria 2* defines the time taken to transfer data during startup (i.e., constructing the distribution table); this time should be less than a certain threshold defined as $T_{acceptance}$. Table 2 clarifies the meaning of symbols used in figure 12. The next step is to model the behavior of system components to acquire the expected services.

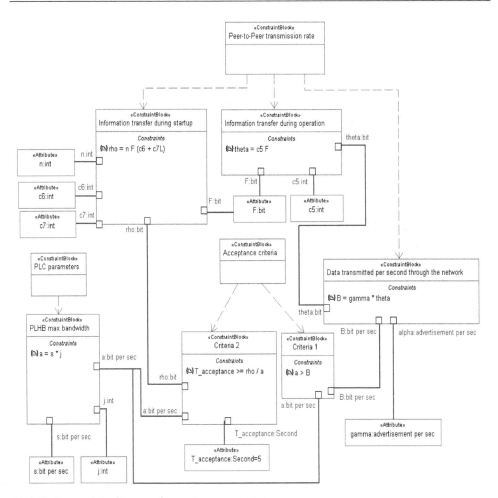

Fig. 12. Parametric diagram for system constraints

3.1.4 Behavior model

The behavior model is aiming at formalizing system behavior, and reconciling it with other requirements. In SysML, behaviors can be represented in different ways; they can be represented by *Activity diagrams*, *Sequence diagrams*, and *State machine diagrams*. We will show how they can give different views for different parts of the system.

Figure 13(a) shows the state machine representing the states of the decentralized technique. When the system startup, the *Construct DT* state is initiated, and each node starts to broadcast the information of its local video contents. Neighboring nodes receive this information and construct their *Distribution Table (DT)*. The DT contains tuples that consist of a unique video file identifier accompanied with the IP address of the node storing this file. When the construction process completes, the *Normal running* state begins, and nodes start to run normally and exchange video contents. The *Update DT* state is fired in two cases. First, when a node

Symbol	Meaning
rho (ρ)	Total amount of information (bytes) transmitted during construction of DT
n	Total number of peers
F	Messages sent between two nodes
$c_5..c_7$	Constants
theta (θ)	Total amount of information (bytes) transmitted during normal operation when one peer advertise one local database change
B	The amount of data per second transmitted through the network
gamma (γ)	Advertisement per second
L	Number of video files stored in a local storage
a	PLHB maximum bandwidth
s	Maximum bandwidth of a single PLB
j	Number of PLBs
$T_{acceptance}$	Maximum delay needed to complete the transmission of ρ or θ

Table 2. Constraints symbols

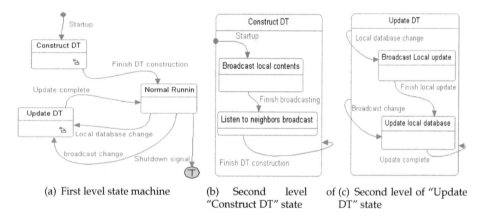

(a) First level state machine   (b) Second level of (c) Second level of "Update "Construct DT" state   DT" state

Fig. 13. State Machine Diagram

has a change in its local video contents, it updates its local DT and broadcasts the change to allow other nodes to update their local DT. Second, when it receives a *broadcast change* from neighboring nodes. When the update process finishes, the *Normal running* state is fired by an *Update complete* signal. The system closes when a shutdown signal is detected. As any SysML diagram, a state can decompose into more detailed levels. This is indicated by a small icon at the right bottom corner of the state. The sub-levels are shown in figuers 13(b) and 13(c) to present a deeper presentation of *Construct DT* and *Update DT* states, respectively.

Each system state has its own *Activity diagram*. It includes the actions needed to fulfill the state, the signals required to initiate the state, and signals to fire a transition to another state.

Figure 14 shows the behavior of state *Update DT*. It is initiated by receiving a signal indicating a change in a local file; then an update of a local database is performed, followed by broadcasting an update signal to neighboring nodes. If a *broadcast change* signal is received, the node checks if it is a new message from a neighbor or it was its own broadcast message. It

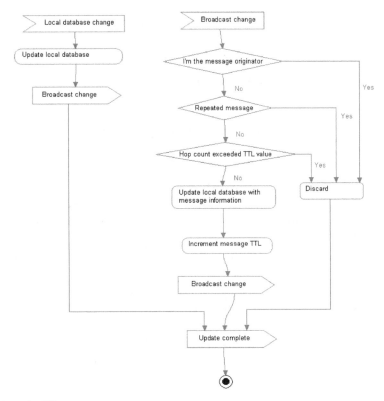

Fig. 14. Activity Diagram

updates its local database, rebroadcast the message, and send an *Update complete* signal to fire a transition to the normal running state.

3.2 System evaluation

We are interested in checking the proposed configuration (i.e., distributed table and PLC network) with the criteria shown in Figure 12 to see if it is feasible or not. According to the values indicated in (Akl et al., 2010) we can calculate the maximum bandwidth supported by each PLHB

$$a = s * j = 3480 * 20 = 69600bit/sec = 0.06638Mb/sec = 8.496KB/sec \qquad (3)$$

As shown in (Loureiro & Anzaloni, 2011), when $\gamma = 20adv/sec$ and $n = 200peers$, then

$$B = 0.1Mb/sec \qquad (4)$$

This can be interpreted as having 200 passengers, where 20 of them are performing an update to their DT. Since we will compare the value of B with the maximum bandwidth of PLHB, then we are assuming the worst case where all advertisements are initiated at the same PLHB segment.

Furthermore, $\rho = 3371.3KB$ at $n = 200$, so we can deduce its value at $n = 20$, where

$$\rho = (3371.3 * 200)/20 = 337.13KB \tag{5}$$

From 3 and 4, we find that $a < B$, so it does not fulfill the first acceptance criteria in Figure 12.

From 3 and 5, we calculate the time (T) required by the PLHB to transfer the data needed to construct DT is

$$T = \rho/a = 337.13/8.4969 = 39.681sec \tag{6}$$

This is not an accepted value because it must be less than $T_{acceptance}$ (i.e., 5 seconds).

Since both criteria are not fulfilled, then we can say that under this configuration, it is not feasible to use the decentralized technique with this PLC network. The available solutions for this problem are:

• To enhance the performance of the decentralized technique to have a less value for B and T

• To enhance the performance of the PLC network to handle more traffic

• To change the value of $T_{acceptance}$ to allow the system to accept more delay.

To achieve these changes, the designer has to change the behaviour diagrams. He also may change or add or remove some components in the block diagrams. Obviously, $T_{acceptance}$ in parametric diagram needs to be changed if the third solution is considered. If possible, some requirements may be altered to minimize the constraints imposed on the design.

3.3 Discussion

IFE is a large system with various components and parameters, especially in an aircraft environment with strict regulations. SysML provides a solution to model and verify such system. The modeling process starts by defining all parties involved with the system and gathering their requirements; this step helps to have a design that complies with their needs. These requirements are presented in a requirement diagram to show consistency, and relations between requirements and constraints. Moreover, it shows system components that are responsible for satisfying the requirements. System components are modeled using block diagrams. The block diagram shows the relations and connections between different components, define the items flowing between them, and the services they provide or need. The behavior of these components is modeled through different diagrams, where each of them represents a different view of the desired behavior. The behavior diagrams show how components can satisfy needed requirements. During the design of these models, parametric diagrams are considered to model system constraints.

The design process life cycle is not a sequential one; this means that at any step, changes can be done to a previous step. For example, during the behaviour diagram design, changes can be done to block or requirement diagrams. However, changes to requirements must be done after the approval of stakeholders. At the end of the design process, all components and behaviours must fulfill all requirements and constraints.

4. Conclusion

Since the very beginning, IFE systems were targeting passenger comfortableness. This target was the main intention to develop services dedicated to passengers. As time goes, business requirements changes, so IFE systems start to reveal another dimension of services to support crew members and airline companies in order to facilitate crew tasks and increase airline revenue. Recent technological advancements helped designers to offer various designs and services. However, this variations increased system complexity and former design techniques become less efficient.

SysML is offered as indispensable tool for modeling complex systems. It can formalize all parts of the system, so that bug tracking, and future enhancements become more manageable. In this work, we showed the design steps for a part of an IFE system and how it can be modeled. Through SysML capabilities, we were able to integrate two different techniques; the Distribution Table for a peer-to-peer network, and the PLC network. These proposals were done by two independent research teams. However, SysML modeling allowed us to verify if these proposals can be used together in the same system or not, and if not, what are the possible available solutions.

5. Future focus areas for IFE systems

IFE systems are still in their development phase and different topics are still under research. In this section, we propose some ideas to be integrated in future designs. Although IFE system development made a great leap in the past years, but there are still various issues that need further research. These developments range from enhancing current systems to adding new components and services. As technology improves, more advanced devices can be used to enhance current components such as increasing network bandwidth, using more accurate contactless sensors, wireless devices, and lighter components. There is no limit for new services that can be added to IFE systems. Nowadays, a passenger who takes different connections to his destination may not be able to continue his selected IFE content even if he is using the same airline. An attractive service is to allow him to continue unfinished IFE content when changing to the next connection, so he can enjoy the selected service for the whole trip regardless of any flight change. Another service is to create a personal profile through which he can customize his favorite contents before taking the flight, so he does not waste time for selecting items during the flight, and his profile can be used for future travels. For health services, automatic pop-up reminders can be used to stop passengers from being stick to the entertainment content. Using 3D displaying devices can introduce a new sensation to IFE entertainment. Furthermore, hologram images can be used to present safety instructions instead of crew members.

6. References

AeroMobile (Last visit 2011). http://www.aeromobile.net/.

Airships.net (Last visit 2011). http://www.airships.net.

Akl, A., Gayraud, T. & Berthou, P. (2010). Investigating Several Wireless Technologies to Build a Heteregeneous Network for the In-Flight Entertainment System Inside an Aircraft Cabin, *The Sixth International Conference on Wireless and Mobile Communications (ICWMC)* pp. 532–537.

Akl, A., Gayraud, T. & Berthou, P. (2011). A New Wireless Architecture for In-Flight Entertainment Systems Inside Aircraft Cabin, *International Journal on Advances in Networks and Services* 4, no. 1 & 2(ISSN 1942-2644): 159–175.

Aksoy, S., Atilgan, E. & Akinci, S. (2003). Airline services marketing by domestic and foreign firms: differences from the customers' viewpoint, *Journal of Air Transport Management* 9(6): 343–351.

Alamdari, F. (1999). Airline in-flight entertainment: the passengers' perspective, *Journal of Air Transport Management* 5(4): 203–209.

AmericanAirlines (2011). http://www.aa.com/i18n/travelInformation/duringFlight/onboa rdTechnology.jsp.

Balcombe, K., Fraser, I. & Harris, L. (2009). Consumer willingness to pay for in-flight service and comfort levels: A choice experiment, *Journal of Air Transport Management* 15(5): 221–226.

Bani-Salameh, Z., Abbas, M., Kabilan, M. K. & Bani-Salameh, L. (2010). Design and Development of Systematic Interactive Multimedia Instruction on Safety Topics for Flight Attendants, *Proceeding of the 5th International Conference on e-Learning* pp. 327–342.

Chang, C.-Y. & Li, S.-T. (2011). Active Noise Control in Headsets by Using a Low-Cost Microcontroller, *IEEE Transactions On Industrial Electronics* 58(5): 1936–1942.

Chang, Y.-H. & Liao, M.-Y. (2009). The effect of aviation safety education on passenger cabin safety awareness, *Safety Science* 47(10): 1337–1345.

Davies, R. & Birtles, P. J. (1999). *Comet - The World's First Jet Airliner*, 1st edn, The Crowood Press Ltd.

DeltaAirline (Last visit 2011). http://www.delta.com/traveling_checkin/inflight_serv-ices/products/wi-fi.jsp.

Diaz, N. R. & Esquitino, J. E. J. (2004). Wideband Channel Characterization for Wireless Communications inside a short haul aircraft, *Vehicular Technology Conference*, pp. 223–228.

ECAB (Last visit 2011). http://ec.europa.eu/research/transport/projects/items/e_ca-b_en.htm.

FlyNet (Last visit 2011). http://konzern.lufthansa.com/en/themen/net.html.

Holzbock, M., Hu, Y.-F., Jahn, A. & Werner, M. (2004). Advances of aeronautical communications in the EU framework, *International Journal of Satellite Communications and Networking* 22(1): 113–137.

Hrasnica, H., Haidine, A. & Lehnert, R. (2004). *Broadband Powerline Communications Networks*, John Wiley & Sons, Ltd.

Hull, E., Jackson, K. & Dick, J. (2011). *Requirements Engineering*, 3rd edn, Springer-Verlag London.

Itzel, L., Tuttlies, V., Schiele, G. & Becker, C. (2010). Consistency Management for Interactive Peer-to-Peer-based Systems, *Proceedings of the 3rd International ICST Conference on Simulation Tools and Techniques*, ICST (Institute for Computer Sciences, Social-Informatics and Telecommunications Engineering), pp. 1–8.

Jahn, A. & Holzbock, M. (2003). Evolution of aeronautical communications for personal and multimedia services, *IEEE Communications Magazine* 41: 36–43.

Khan, A. M. (2010). *Communication Abstraction for Data Synchronization in Distributed Virtual Environments Application to Multiplayer Games on Mobile Phones*, PhD thesis, Université d'Evry-Val d'Essonne.

Khan, A. M., Arsov, I., Preda, M., Chabridon, S. & Beugnard, A. (2010). Adaptable Client-Server Architecture for Mobile Multiplayer Games, *Proceedings of the 3rd International ICST Conference on Simulation Tools and Techniques* 11: 1–7.

Kim, J., Choi, J., Chang, D., Kwon, T., Choi, Y. & Yuk, E. (2005). Traffic Characteristics of a Massively Multi-player Online Role Playing Game, *Proceedings of 4th ACM SIGCOMM workshop on Network and system support for games*, ACM, pp. 1–8.

Lansford, J., Stephens, A. & Nevo, R. (2001). Wi-Fi (802.11b) and Bluetooth: Enabling Coexistence, *IEEE Communications Magazine* pp. 20–27.

Leavitt, N. (2007). For Wireless USB, the Future Starts Now, *IEEE Computer Society* 40(7): 14–16.

Liou, J. J. (2011). Consumer attitudes toward in-flight shopping, *Journal of Air Transport Management* 17(4): 221–223.

Liu, H. (2007). In-Flight Entertainment System: State of the Art and Research Directions, *Second International Workshop on Semantic Media Adaptation and Personalization (SMAP 2007)*, Second International Workshop on Semantic Media Adaptation and Personalization (SMAP 2007), pp. 241–244.

Liu, H. & Rauterberg, M. (2007). Context-aware In-flight Entertainment System, *Proceedings of Posters at HCI International* pp. 1249–1254.

Loureiro, R. Z. & Anzaloni, A. (2011). Searching Content on Peer-to-Peer Networks for In-Flight Entertainment, *IEEE Aerospace conference* pp. 1–4.

Lufthansa (2011). http://www.lhsystems.com/solutions/infrastructure-services/wireless-in-flight-entertainment.htm.

Moraitis, N., Constantinou, P., Fontan, F. P. & Valtr, P. (2009). Propagation Measurements and Comparison with EM Techniques for In-Cabin Wireless Networks, *Journal EURASIP Journal on Wireless Communications and Networking - Special issue on advances in propagation modelling for wireless systems* 5: 1–13.

Nadadur, G. & Parkinson, M. B. (2009). Using designing for human variability to optimize aircraft seat layout, *SAE International Journal of Passenger Cars-Mechanical Systems* 2: 1641–1648.

Niebla, C. (2003). Coverage and capacity planning for aircraft in-cabin wireless heterogeneous networks, *IEEE Vehicular Technology Conference* pp. 1658–1662.

Qantas (2011). http://www.qantas.com.au/travel/airlines/inflight-communications/global/en.

Qi, H., Malone, D. & Botvich, D. (2009). 802 . 11 Wireless LAN Multiplayer Game Capacity and Optimization, *8th Annual Workshop on Network and Systems Support for Games (NetGames)* pp. 1–6.

Radzik, J., Pirovano, A., Tao, N. & Bousquet, M. (2008). Satellite system performance assessment for In-Flight Entertainment and Air Traffic Control, *Journal of Space Communication* 21: 69–82.

ROW44 (2011). http://row44.com/products-services/broadband/.

Schumm, J., Setz, C., Bächlin, M., Bächler, M., Arnrich, B. & Tröster, G. (2010). Unobtrusive physiological monitoring in an airplane seat, *Personal and Ubiquitous Computing* 14(6): 541–550.

Sohn, J. M., Baek, S. H. & Huh, J. D. (2008). Design issues towards a high performance wireless USB device, *IEEE International Conference on Ultra-Wideband* 3: 109–112.

Tan, C., Chen, W., Verbunt, M., Bartneck, C. & Rauterberg, M. (2009). Adaptive Posture Advisory System for Spinal Cord Injury Patient, *Proceedings of the ASME International Design Engineering Technical Conferences & Computers and Information in Engineering Conference IDETC/CIE* pp. 1–7.

Tan, C. F., Iaeng, M., Chen, W., Kimman, F. & Rauterberg, G. W. M. (2009). Sleeping Posture Analysis of Economy Class Aircraft Seat, *Proceedings of the World Congress on Engineering (WCE)* 1: 532–535.

Thales (2011). http://www.thalesgroup.com/Case_Studies/Markets/Aerospace/Inno-vating _for_inflight_entertainment_(IFE)/?pid=10295.

Thompson, H. (2004). Wireless and Internet communications technologies for monitoring and control, *Control Engineering Practice* 12(6): 781–791.

Udar, N., Kant, K., Viswanathan, R. & Cheung, D. (2007). Characterization of Ultra Wide Band Communications in Data Center Environments, *Procceedings of ICUWB* pp. 322–328.

Vink, P. (2011). *Aircraft Interior Comfort and Design*, 1st edn, CRC Press (Taylor and Francis Group).

Westelaken, R., Hu, J., Liu, H. & Rauterberg, M. (2010). Embedding gesture recognition into airplane seats for in-flight entertainment, *Journal of Ambient Intelligence and Humanized Computing* 2(2): 103–112.

Young, R. R. (2004). *The requirements engineering handbook*, 1 edn, Artech House Inc.

Youssef, M., Vahala, L. & Beggs, J. (2004). Wireless network simulation in aircraft cabins, *IEEE Antennas and Propagation Society Symposium* 3: 2223–2226.

4

Methods for Analyzing the Reliability of Electrical Systems Used Inside Aircrafts

Nicolae Jula[1] and Cepisca Costin[2]
[1]Military Technical Academy of Bucharest
[2]University Politehnica of Bucharest
Romania

1. Introduction

This chapter presents two solutions to perform reliability analysis of electrical systems installed on aircrafts. The first method for determining the reliability of electrical networks is based on an analogy between electrical impedance and reliability. The second method is based on application of Boolean algebra to the study of reliability in electrical circuits. By using these research methods we obtain information on operational safety of the electrical systems on board of an airplane, either for the entire system or for each of its components (Jula, 1986). The results allow further optimization of the construction of electrical system used on aircrafts (Aron et al., 1980), (Jula et al., 2008).

2. Calculating electrical impedance and reliability – an analogy

Establishing the reliability of structures resulting from the analysis of electrical systems installed on board of aircrafts can be achieved by direct calculations, but involves a long working time as a result of taking into account all possible situations that can occur during system operation (Reus, 1971), (Hoang Pham ,2003), (Levitin, G. et al., 1997).

A more efficient calculation method for complex structures can be achieved by applying equivalent transformation methods in terms of reliability, similar to the transformation theorems for electrical circuits applied to determine the equivalent impedance between two nodes (Moisil, 1979), (Drujinin,1977), (Billinton, 1996).

2.1 Short presentation of the analogy method

To highlight the approximations introduced by this method of calculation consider a group of elements connected in series, with the likelihood of downtime q_1, q_2, ..., q_n. Using transformation theorem for elements in series, these elements can be replaced with a resultant, a single item that has a probability of downtime q, (Drujinin,1977), given by:

- The exact formula

$$q = 1 - \prod_{i=1}^{n}(1 - q_i) \tag{1}$$

- The approximation of order 1

$$q = \sum_{i=1}^{n} q_i \tag{2}$$

- The approximation of order 2

$$q = \sum_{i=1}^{n} q_i - \frac{1}{2}\sum_{i=1}^{n}\sum_{j=1}^{n} q_i q_j \tag{3}$$

For the approximation of order 1, the error made is of the order of magnitude q_i^2, while for 2nd order the approximation error is q_i^3, etc.

Therefore for order 1 approximation, the probabilities of downtimes $q_1, q_2, ..., q_n$ of elements connected in series are added together as if determining the equivalent impedance of a circuit with electrical components connected in series.

A group of elements connected in parallel with the probability of downtimes $q_1, q_2, ..., q_n$ can be replaced by one single element that has a probability of downtime:

$$q = \prod_{i=1}^{n} q_i \tag{4}$$

In this case, the equivalent probability of downtime is achieved as a product of individual probabilities; therefore the result in this case is different from the equivalent impedance of an electrical circuit made of components in parallel.

A group of elements with delta connection, with the likelihood of downtime q_{12}, q_{23}, q_{31} may be replaced by another group of elements connected in star with the probability of downtime q_1, q_2, q_3. The relations for transformation are:

$$\begin{aligned} q_1 &= q_{12}q_{31} \\ q_2 &= q_{23}q_{12} \\ q_3 &= q_{31}q_{23} \end{aligned} \tag{5}$$

with an approximation error proportional with $q_{12} \cdot q_{23} \cdot q_{31}$.

Relation (5) was deducted under the assumption that the reliability of the circuit between two points, for example between point 1 and point 2 - Figure 1 - is the same for both connections in two borderline cases, namely:

- The third point is offline,
- The third point is connected to one of the first two.

Under these conditions the following relationships are obtained:

$$\begin{aligned} q_1 + q_2 - q_1 q_2 &= q_{12}(q_{23} + q_{31} - q_{23}q_{31}) \\ q_2 + q_3 - q_2 q_3 &= q_{23}(q_{31} + q_{12} - q_{31}q_{12}) \\ q_3 + q_1 - q_3 q_1 &= q_{31}(q_{12} + q_{23} - q_{12}q_{23}) \end{aligned} \tag{6}$$

$$q_1 + q_{23} - q_1 q_2 q_3 = q_{12} q_{31}$$
$$q_2 + q_{31} - q_2 q_3 q_1 = q_{23} q_{12} \tag{7}$$
$$q_3 + q_{12} - q_3 q_1 q_2 = q_{31} q_{23}$$

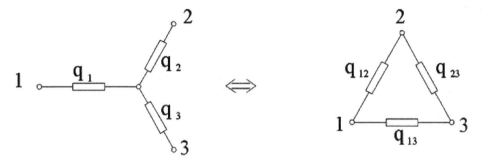

Fig. 1. Star-Delta and Delta - Star transformation for reliability.

It can be seen that the two systems described in (6) and (7) are incompatible. But if you take into account that the components used in electrical circuits on board of an aircraft are characterized by $q \ll 1$, approximate solutions can be utilized (Aron & Paun, 1980).

Neglecting the smaller higher-order terms of the transformation delta-star, in this case the third order component, equations in (6) become:

$$q_1 + q_2 = q_{12} q_{23} + q_{12} q_{31}$$
$$q_2 + q_3 = q_{23} q_{31} + q_{23} q_{12} \tag{8}$$
$$q_3 + q_1 = q_{31} q_{12} + q_{31} q_{23}$$

If the second equation is multiplied by (-1) and all the system equations are added, equation (9) is obtained:

$$q_1 = q_{12} q_{31} \tag{9}$$

Applying the same methodology for the other two remaining equations in (8) results the below equivalence for delta-star transformation:

$$q_1 = q_{12} q_{31}$$
$$q_2 = q_{23} q_{12} \tag{10}$$
$$q_3 = q_{31} q_{23}$$

From (7) and using the same methodology, relationships for star-delta transformation are obtained (Hohan, 1982):

$$q_{12} = \sqrt{\frac{q_1 q_2}{q_3}} \quad q_{23} = \sqrt{\frac{q_2 q_3}{q_1}} \quad q_{31} = \sqrt{\frac{q_3 q_1}{q_1}} \tag{11}$$

2.2 The analogy method applied for electrical circuits used in aircrafts

Example 1. The diagram presented in Figure 2.a corresponds to a three-phase electrical generator, part of the airplane power system, powered by a three-phase electric motor, both having their stators with delta connection. The transformed version of the diagram according to the analogy method is shown in Figure 2.b.

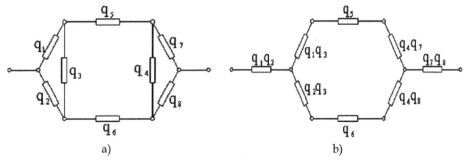

a)   b)

Fig. 2. Delta-star transformation – example 1.

The transformation delta – star applied to q_1, q_2, q_3 and q_4, q_7, q_8 becomes a simple network configuration for which downtime can be established with the specific probability when applying the previously derived relations:

$$Q = q_1q_2 + q_7q_8 + (q_1q_3 + q_5 + q_4q_7)(q_2q_3 + q_6 + q_4q_8)$$

$$Q = q_1q_2 + q_5q_6 + q_7q_8 + q_1q_3q_6 + q_1q_3q_5 + q_4q_6q_7 + q_4q_5q_8$$

If the components have the same probability q, then the probability of downtime Q is:

$$Q = 3q^2 + 4q^3$$

Example 2. Figure 3 shows the diagram of a measurement instrument based on logometric principle, used to measure engine temperature or quantity of existing fuel in the plane tanks (Jula, 1986).

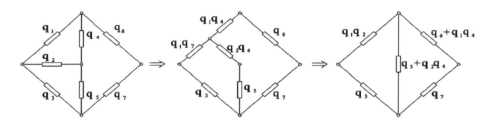

Fig. 3. Transformations for the measurement instrument – example 2.

The relations obtained for the probability of downtime Q after two transformations are:

$$Q = q_1 q_2 q_3 + q_7 (q_6 + q_1 q_4) + q_1 q_2 q_7 (q_5 + q_2 q_4) + q_3 (q_6 + q_1 q_4)(q_5 + q_2 q_4)$$

$$Q \cong q_6 q_7 + q_1 q_2 q_3 + q_1 q_4 q_7 + q_3 q_5 q_6$$

If the components have the same probability of downtime q, it results:

$$Q \cong q^2 + 3q^3$$

Example 3. The diagram in Figure 4 corresponds to an aircraft specific electromagnetic system powered by multiple nodes.

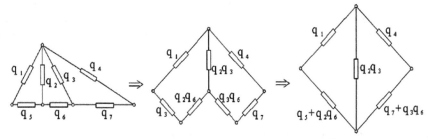

Fig. 4. Successive transformation of the electromagnetic system – example 3.

The downtime probability Q, resulting from the transformations illustrated above is:

$$Q = q_1 (q_5 + q_2 q_6) + q_4 (q_7 + q_3 q_6) + q_1 q_2 q_3 (q_7 + q_3 q_6) + q_7 + q_2 q_3 q_4 (q_5 + q_2 q_6)$$

$$Q \cong q_1 q_5 + q_4 q_7 + q_1 q_2 q_6 + q_3 q_4 q_6$$

If the components have the same probability q of downtime, it results:

$$Q \cong 2q^2 + 3q^3$$

Alternatively, a more efficient transformation is presented in Figure 5.

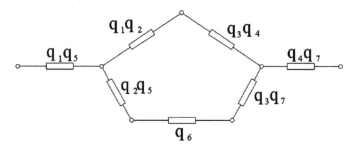

Fig. 5. A version of the final state after the transformation.

A relation for this state is:

$$Q = q_1 q_5 + q_4 q_7 + (q_6 + q_2 q_5 + q_3 q_7)(q_1 q_2 + q_3 q_4)$$

$$Q \cong q_1 q_5 + q_4 q_7 + q_1 q_2 q_6 + q_1 q_4 q_6$$

Whereby the result is identical to the one previously obtained, the calculation time is significantly reduced.

2.3 Conclusions regarding the analogy method

The method draws on the similarity between the calculus for the electrical impedance and the reliability one, allowing the use of simple relationships and reducing the number of equations to be solved. In case of complex networks other methods would lead to difficulties in obtaining results in short time, while the analogy method, with its rather low number of calculations ensures a time efficient way of finding the downtime probability of any electrical circuit.

If one or more circuit elements are less reliable than other parts of the circuit, and therefore its downtime probability is high, the transformation can get more accurate approximations of the real state of the system than other methods, mainly due to the multiplier effect contained.

3. The method based on Boolean logical structures

Large-scale systems reliability analysis is based on the quantification of the failure process at the structural level. Thus, any system downtime is a result of a quantified sequence of states in the failure process. The quantification level can be chosen in accordance with the desired goal and probability, down even to individual components of the system. The more detailed the quantification, the more accurate would be the resulting probability (Reus, 1971) (Muzi, 2008).

The conceptual representation of an emergent downtime is formed by a series of primary events, interconnected through different Boolean logical structures, which indicate the possible combinations of those elements having as result a system failure (Denis-Papin& Malgrange, 1970), (Chern & Jan, 1986). Thus determining the reliability of an aircraft electrical system using Boolean algebra actually means calculating the probability of a "failure" event.

3.1 Principles of the Boolean method

From the structural point of view, for the reliability analysis, we will use the terms:

-   Primary elements – components or blocks at the base level of the quantification,
-   Primary failures – primary elements failures,
-   Unwanted event – system failure state,
-   Failure mode – the set of primary elements that when simultaneously in failure mode, drives to a system failure
-   Minimal failure mode – the smallest set of primary components that when simultaneously in failure mode, drive to a system failure
-   Hierarchic level – all elements that are structurally equivalent and having equivalent positions in the system failure representation.

The method is based on binary logic. Thus, a system function is equivalent to a binary function, which variables are the events (the failures).

This binary function:

$$Y = f\left(X_1, X_2, ..., X_n\right) \tag{12}$$

is synthesized with logical elements AND/OR, using the following symbols and states:

- $\bigcup$ (Reunion) for the function OR
- $\bigcap$ (Intersection) for the function AND

X_i is 1 if the primary element is good and 0 otherwise, and Y is 1 if the system is good and 0 otherwise.

The method representation is depicted in Figure 6. For the reliability function indicators calculus, in the hypothesis of the failure intensity having an exponential distribution, we use the relations:

$$R(t) = \exp\left(-\sum_{i=1}^{n} \lambda_i t\right) = \exp(-\wedge t) \tag{13}$$

$$R(t) = 1 - \prod_{i=1}^{n}\left[1 - \exp(\lambda_i t)\right] \tag{14}$$

where: $\wedge = \sum_{i=1}^{n} \lambda_i$.

Relation (13) is used for the serial connection and relation (14) is used for the parallel connection of the elements.

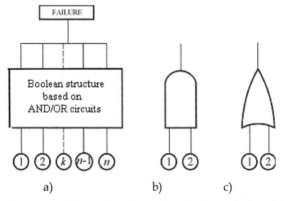

Fig. 6. a) The general concept of the method based on Boolean algebra (1, 2,..., n are independent primary events); b) the schematics of the logic function AND; c) the schematics of the logic function OR.

3.2 Method application for determining the reliability of the aircrafts electric circuits

In order to exemplify the method for the reliability indicators determination, we will focus on the DC electrical power supply system of an aircraft. Figure 7 depicts the electric power supply system of an aircraft.

In principle, this electric power supply system is present (as the main electric power supply system) in a large number of military aircrafts ranging from the MiG family (21, 23, 27, 29,31,35), Su (30,33,34,35,37) to Chengdu (J-10), Shenyang (J-11) and ORAO. The example refers only to a DC electric power supply system nevertheless the method can be used in alternative current and mixed systems set-ups. In Figure 7:

- 1E – starter-generator – startup time of several seconds (as a starter), after a successful start (three attempts permitted) it goes to a generator regime, supplying a 28V DC voltage
- 4E – accumulator switch
- 5E – inverse polarity protection diode
- 13E – accumulator
- 14E – accumulator to DC bar switch
- 24E – generator to DC bar coupler / de-coupler
- 47E – fuse
- 27E – voltage regulator.

The emerging failure state diagram using AND/OR elements is depicted in Figure 8. The failure event is the loss of voltage at the 28V bar.

For the failure intensity λ_i of the components we use the relation:

$$\lambda_i = k\lambda_0 \qquad (15)$$

where: k – maintenance and way-of-use coefficient (for aircraft components the coefficient varies between 120 and 160); λ_0 – failure intensity – manufacturer specific data.

The data relative to the electric power supply system are presented in Table 1.

Symbol	Description	$\lambda_0 \left[h^{-1} \right]$	No.	k	$\lambda_i = nk\lambda_0 \left[h^{-1} \right]$	$F_i = 1 - e^{-\lambda_i t}$
4E	Switch	$0.12 \cdot 10^{-6}$	1	160	$\lambda_1 = 1.92 \cdot 10^{-5}$	$F_1 = 1 - e^{-1.92 \cdot 10^{-5} t}$
5E	Diode	$0.6 \cdot 10^{-6}$	1	160	$\lambda_1 = 9.6 \cdot 10^{-5}$	$F_2 = 1 - e^{-9.6 \cdot 10^{-5} t}$
13E	Accumulator	$1.4 \cdot 10^{-6}$	1	160	$\lambda_1 = 22.4 \cdot 10^{-5}$	$F_3 = 1 - e^{-22.4 \cdot 10^{-5} t}$
14E	Coupler	$0.4 \cdot 10^{-6}$	1	160	$\lambda_1 = 6.4 \cdot 10^{-5}$	$F_4 = 1 - e^{-6.4 \cdot 10^{-5} t}$
47E	Fuse	$2.75 \cdot 10^{-6}$	1	160	$\lambda_1 = 44 \cdot 10^{-5}$	$F_5 = 1 - e^{-44 \cdot 10^{-5} t}$
-	Contacts 1	$0.1 \cdot 10^{-6}$	1	160	$\lambda_1 = 16 \cdot 10^{-5}$	$F_6 = 1 - e^{-16 \cdot 10^{-5} t}$

Table 1. Part I

Symbol	Description	$\lambda_0\left[\text{h}^{-1}\right]$	No.	k	$\lambda_i = nk\lambda_0\left[\text{h}^{-1}\right]$	$F_i = 1 - e^{-\lambda_i t}$
1E	Starter-generator	$6 \cdot 10^{-6}$	1	160	$\lambda_1 = 96 \cdot 10^{-5}$	$F_8 = 1 - e^{-96 \cdot 10^{-5} t}$
24E	Coupler / Decoupler	$0.25 \cdot 10^{-6}$	1	160	$\lambda_1 = 4 \cdot 10^{-5}$	$F_9 = 1 - e^{-4 \cdot 10^{-5} t}$
27E	Voltage regulator	$13 \cdot 10^{-6}$	1	160	$\lambda_1 = 208 \cdot 10^{-5}$	$F_{10} = 1 - e^{-208 \cdot 10^{-5} t}$
-	Contacts 1	$0.1 \cdot 10^{-6}$	1	160	$\lambda_1 = 16 \cdot 10^{-5}$	$F_{11} = 1 - e^{-16 \cdot 10^{-5} t}$

Table 1. Part II

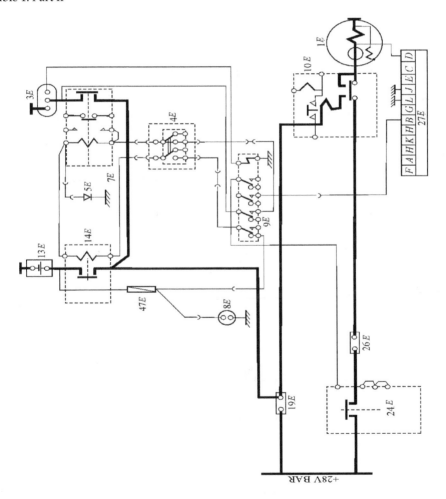

Fig. 7. The electric power supply diagram for a DC main electric supply system aircraft (fragment).

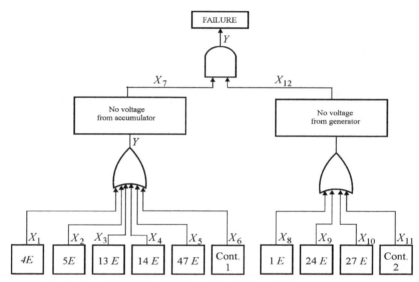

Fig. 8. The logic structure that drives to the system failure status.

In these conditions, the Boolean function associated to the logic structure depicted in Figure 8 has the following form:

$$Y = X_7 \cap X_{12} = (X_1 \cup X_2 \cup X_3 \cup X_4 \cup X_5 \cup X_6) \cap (X_8 \cup X_9 \cup X_{10} \cup X_{11}) \tag{16}$$

To transform the logic equation into algebraic form we use the following relations

$$X_1 \cap X_2 = X_1 \cdot X_2 \; ; \; X_1 \cup X_2 = X_1 + X_2 - X_1 X_2 \; ; \; \bigcup_{i=1}^{n} X_i = 1 - \prod_{i=1}^{n}(1 - X_i) \tag{17}$$

Thus, we have

$$Y = \left[1 - (1 - X_1)(1 - X_2)(1 - X_3)(1 - X_4)(1 - X_5)(1 - X_6) \right] \cdot \left[1 - (1 - X_8)(1 - X_9)(1 - X_{10})(1 - X_{11}) \right] \tag{18}$$

which is similar to

$$Y = X_7 \cdot X_{12} = \left[1 - \prod_{i=1}^{6}(1 - X_i) \right] \cdot \left[1 - \prod_{k=8}^{11}(1 - X_k) \right] \tag{19}$$

Considering the failure intensity as exponential distribution, the system failure probability is given by the following relations:

$$F(t) = \left\{ 1 - \exp\left[-(\lambda_1 + \lambda_2 + \lambda_3 + \lambda_4 + \lambda_5 + \lambda_6)t \right] \right\} \cdot \left[1 - \exp(-\lambda_8 - \lambda_9 - \lambda_{10} - \lambda_{11})t \right] =$$

$$= 1 - \exp\left[-\sum_{i=8}^{11}\lambda_i t \right] - \exp\left[-\sum_{k=1}^{6}\lambda_k t \right] + \exp\left[-\sum_{\substack{p=1 \\ p \neq 7}}^{11}\lambda_p t \right] \tag{20}$$

$$R(t) = 1 - F(t) = \exp\left[-\sum_{i=8}^{11} \lambda_i t\right] + \exp\left[-\sum_{k=1}^{6} \lambda_k t\right] - \exp\left[-\sum_{\substack{p=1 \\ p \neq 7}}^{11} \lambda_p t\right] \tag{21}$$

$$MTBF = \int_0^\infty R(t)\,dt = \frac{1}{\sum\limits_{i=8}^{11} \lambda_i t} + \frac{1}{\sum\limits_{k=1}^{6} \lambda_k t} - \frac{1}{\sum\limits_{\substack{p=1 \\ p \neq 7}}^{11} \lambda_p t} =$$

$$= \frac{1}{(96 + 4 + 208 + 16)\cdot 10^{-5}} + \frac{1}{(1.92 + 9.6 + 22.4 + 6.4 + 44 + 16)\cdot 10^{-5}} +$$

$$+ \frac{1}{(1.92 + 9.6 + 22.4 + 6.4 + 44 + 16 + 96 + 4 + 208 + 16)\cdot 10^{-5}}$$

On results $MTBF$ = 1069.79 hours.

Thus, mean time between failures in the non improved system may be approximated as follows $MTBF \cong 1070$ hours.

3.3 Reliability optimization of electric power supply in the aircraft industry

We can improve the electric power supply system reliability using a redundant (reserve) subsystem. The proposed improved electric power supply, including the back-up subsystem (dotted lines) is depicted in Figure 9.

Further on we will analyze the improved electric power supply system reliability, using the Boolean method presented in chapter 3.2. This analysis also allows a determination of a relation between the system reliability and the system weight. Such a relation is useful when emphasizing the variation of the system reliability with the total weight of system components.

Through a compared analysis of different reliability improving variants, imposing as minimum condition the component weight, we can obtain an optimal solution. The logic structure that drives to the system failure status (for the improved system schematics) is depicted in Figure 10.

Table 2 presents the values of the failure intensity for the supplementary components from the back-up system, in the exponential distribution hypothesis.

Symbol	Description	$\lambda_0\ [h^{-1}]$	No.	k	$\lambda_i = nk\lambda_0\ [h^{-1}]$	$F_i = 1 - e^{-\lambda_i t}$
60E	Coupler	$0.4 \cdot 10^{-6}$	1	160	$\lambda_1 = 6.4 \cdot 10^{-5}$	$F_1 = 1 - e^{-6.4\cdot 10^{-5} t}$
61E	Switch	$0.12 \cdot 10^{-6}$	1	160	$\lambda_1 = 1.92 \cdot 10^{-5}$	$F_2 = 1 - e^{-1.92\cdot 10^{-5} t}$
-	Contacts 3	$0.1 \cdot 10^{-6}$	4	160	$\lambda_1 = 6.4 \cdot 10^{-5}$	$F_3 = 1 - e^{-6.4\cdot 10^{-5} t}$

Table 2.

The Boolean function in this case is:

$$Y = (X_{16} \cap X_7) \cap X_{12} = (X_{13} \cup X_{14} \cup X_{15}) \cap$$
$$\cap (X_1 \cup X_2 \cup X_3 \cup X_4 \cup X_5 \cup X_6) \cap \qquad (22)$$
$$\cap (X_8 \cup X_9 \cup X_{10} \cup X_{11}).$$

Transforming in algebraic form, we have:

$$Y = [1 - (1 - X_{13})(1 - X_{14})(1 - X_{15})] \cdot [1 - (1 - X_1)(1 - X_2)(1 - X_3)(1 - X_4)(1 - X_5)(1 - X_6)] \cdot$$
$$\cdot [1 - (1 - X_8)(1 - X_9)(1 - X_{10})(1 - X_{11})] \qquad (23)$$

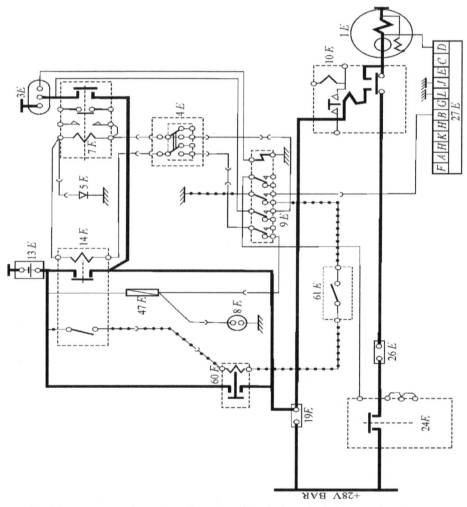

Fig. 9. Electric power supply system of an aircraft including the back-up subsystem (fragment).

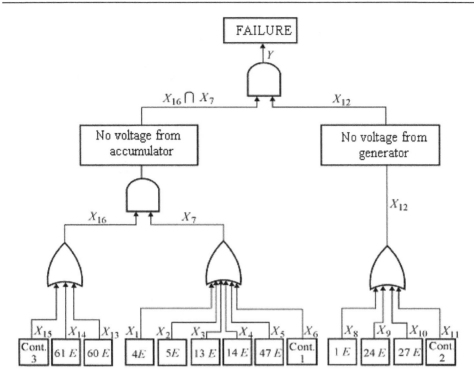

Fig. 10. The logic structure of the electric system presented in fig. 9.

$$Y = \left[1 - \prod_{i=13}^{15}(1 - X_i)\right] \cdot \left[1 - \prod_{k=1}^{6}(1 - X_k)\right] \cdot \left[1 - \prod_{p=8}^{11}(1 - X_p)\right] \qquad (24)$$

From (24) we can determine the system failure probability $F(t)$:

$$F(t) = \left[1 - \exp\left(-\sum_{i=13}^{15}\lambda_i t\right)\right] \cdot \left[1 - \exp\left(-\sum_{k=1}^{6}\lambda_k t\right)\right] \cdot \left[1 - \exp\left(-\sum_{p=8}^{11}\lambda_p t\right)\right] = 1 - \exp\left(-\sum_{i=13}^{15}\lambda_i t\right) -$$

$$- \exp\left(-\sum_{k=1}^{6}\lambda_k t\right) - \exp\left(-\sum_{p=8}^{11}\lambda_p t\right) + \exp\left(-\sum_{\substack{i=1\\i\neq7}}^{11}\lambda_i t\right) + \qquad (25)$$

$$+ \exp\left(-\sum_{\substack{i=1\\i\neq7,8,9,10,11,12}}^{15}\lambda_i t\right) - \exp\left(-\sum_{\substack{i=1\\i\neq7\\i\neq12}}^{15}\lambda_i t\right) + \exp\left(-\sum_{\substack{i=8\\i\neq12}}^{15}\lambda_i t\right)$$

$F(t)$ and $R(t)$ are complementary functions, thus, for the electric power supply system reliability $R(t)$ we will have the following relation:

$$R(t) = \exp\left(-\sum_{i=13}^{15} \lambda_i t\right) + \exp\left(-\sum_{k=1}^{6} \lambda_k t\right) +$$

$$+ \exp\left(-\sum_{p=8}^{11} \lambda_p t\right) - \exp\left(-\sum_{\substack{i=1 \\ i\neq 7}}^{11} \lambda_i t\right) - \qquad (26)$$

$$- \exp\left(-\sum_{\substack{i=1 \\ i\neq 7,8,9,10,11,12}}^{15} \lambda_i t\right) + \exp\left(-\sum_{\substack{i=1 \\ i\neq 7 \\ i\neq 12}}^{15} \lambda_i t\right) - - \exp\left(-\sum_{\substack{i=8 \\ i\neq 12}}^{15} \lambda_i t\right)$$

$$MTBF = \int_{0}^{\infty} R(t)\,dt = \frac{1}{\displaystyle\sum_{i=13}^{15} \lambda_i} + \frac{1}{\displaystyle\sum_{k=1}^{6} \lambda_k} + \frac{1}{\displaystyle\sum_{p=8}^{11} \lambda_p} -$$

$$- \frac{1}{\displaystyle\sum_{\substack{i=1 \\ i\neq 7}}^{11} \lambda_i} - \frac{1}{\displaystyle\sum_{\substack{i=1 \\ i\neq 7,8,9,10,11,12}}^{15} \lambda_i} + \frac{1}{\displaystyle\sum_{\substack{i=1 \\ i\neq 7 \\ i\neq 12}}^{15} \lambda_i} - \frac{1}{\displaystyle\sum_{\substack{i=8 \\ i\neq 12}}^{15} \lambda_i} \cong 6926 \text{ hours} \qquad (27)$$

3.4 Influence of the maintenance and way-of-use coefficient k on $MTBF$

Taking into account the characteristics of the system failure probability - $F(t)$ and reliability $R(t)$ as in Figure 7 and 9, a simulation was made using a Matlab program (Jula et. Al., 2008), which presents the time evolutions of the variables.

Coefficient k from the equation (15) has the starting value $k = 160$. For this value MTBF was calculated both for the initial and the improved systems. The Matlab program helps conduct a complex analysis of the influence of coefficient k on system failure's probability, its reliability and $MTBF$.

Time characteristics $F(t)$ and $R(t)$, for different values of coefficient k are presented below ($k = 120$ (blue), $k = 130$ (red), $k = 140$ (black), $k = 150$ (magenta) and $k = 130$ (green)).

Figures 11 to 13 present the results for the initial system. As it can be seen, the increase of k is directly proportional with function $F(t)$ and inversely proportional with the reliability function $R(t)$. Mean time between failure ($MTBF$) is bigger for small values of the coefficient k.

The same analysis will be conducted for the improved system, in order to compare results. The graphic characteristics are the presented in Figures 14 to 16, while the obtained values both for initial system and improved system are presented in Table 3.

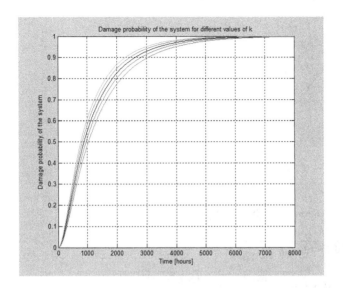

Fig. 11. System failure probability $F(t)$ for different values of k (initial system).

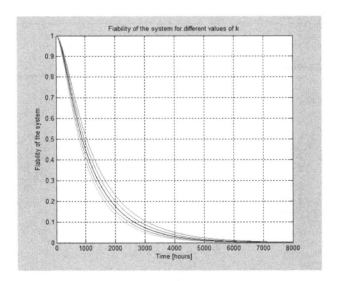

Fig. 12. System's reliability $R(t)$ for different values of k (initial system).

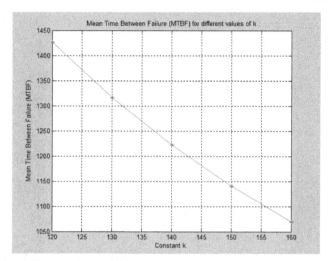

Fig. 13. *MTBF* for different values of *k* (initial system).

MTBF for different k	$k = 120$	$k = 130$	$k = 140$	$k = 150$	$k = 160$
Initial system (fig.3)	1426.4 hours	1316.7 hours	1222.6 hours	1141.1 hours	1069.8 hours
Improved system (fig.4)	9.2354 hours	8.5250 hours	7.9160 hours	7.3883 hours	6.9265 hours
$\gamma = \dfrac{(MTBF)_r}{(MTBF)_0}$	6.4746	6.4745	6.4747	6.4747	6.4746

Table 3.

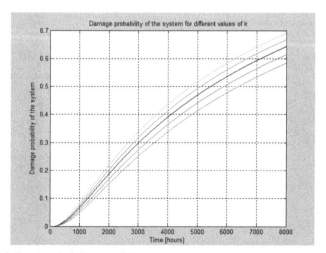

Fig. 14. System failure probability for different values of *k* (improved system).

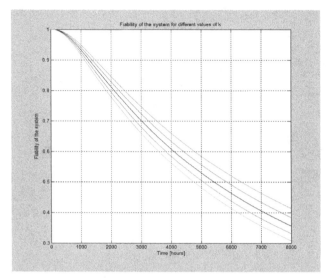

Fig. 15. System's reliability for different values of k (improved system).

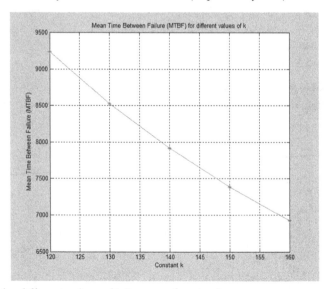

Fig. 16. $MTBF$ for different values of k (improved system).

A comparative presentation of the two systems' reliability for different values of k is depicted in Figure 17 (for initial system with blue lines and red for the improved system).

For the five analyzed values of coefficient k, the improved electric supply with a redundant (reserve) subsystem is characterized by superior values of $MTBF$ compared to the initial system (fig.18).

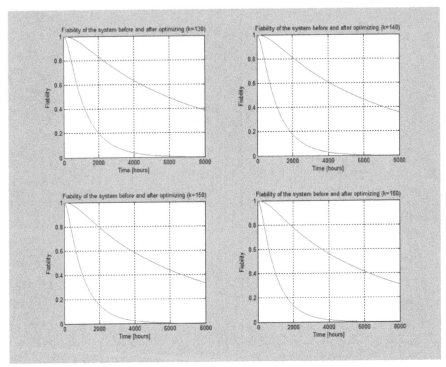

Fig. 17. Comparative analysis of the two systems' reliability for different values of k.

In Figure 18 the evolution of $MTBF$ for the initial system is represented by a dashed line, while the evolution of $MTBF$ for the improved system is represented by a continuous line.

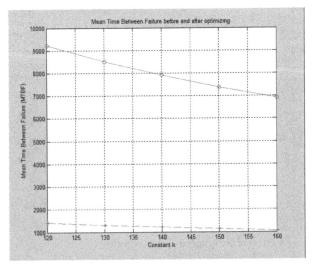

Fig. 18. Evolutions of $MTBF$ for the two systems.

3.5 Conclusions regarding the Boolean method

From the analyzed examples and then results obtained for MTBF, we can conclude that the method can be successfully used in the aircraft industry for determining the reliability of the electrical systems. The *MTBF* influencing parameters in the main system nodes (power supply bars and distribution panels) can be calculated and compared.

Through the failure related logic function analysis we can determine the circuits that can improve the system reliability. In the case presented, through the introduction of the components 60E, 61E and corresponding contacts, substantial increase of the reliability (approximately 6 times higher) was obtained for the 28V DC power supply bar.

We have conducted a complex analysis of the influence of the maintenance and way-of-use coefficient k on system failure probability, system's reliability and *MTBF*.

4. References

Jula, N. (1986). Contribuții la optimizarea circuitelor electrice de la bordul avioanelor militare. PhD Thesis, București, Romania

Moisil, G. (1979). *Teoria algebrică a mecanismelor automate*. Ed. Tehnică, Bucharest

Drujinin, C.V. (1977). *Nadejnot aftometizirovannijh - Sistem*, Energhia, Moskva

Aron, I.; Păun, V. (1980). *Echipamentul electric al aeronavelor*, Editura Didactică și Pedagogică, București, Romania

Hoang Pham (2003). *Handbook of Reliability Engineering*. Springer Verlag

Mathur ,F.P.; De Sousa, P.T.Reliability modeling and analysis of general modular redundant systems, *IEEE Trans.Reliab*. 1975, 24, 296-9

Hohan, I. (1982). *Fiabilitatea sistemelor mari*, E.D.P., Bucharest, Romania

Gnedenko, B.; (1995). *Probabilistic reliability engineering*. New York, John Wiley & Sons

Reus, I. (1971). *Tratarea simbolică a schemelor de comutație*. Ed. Academiei, Bucharest

Muzi, F. Real-time Voltage Control to Improve Automation and Quality in Power Distribution. *WSEAS Transactions on Circuits and Systems*, Issue 6, Vol. 7, 2008

Levitin, G.;Lisnianski, A.; Ben Haim, H.; Elmakis, D. Redundancy optimization for series-paralell multi-state systems. *IEEE Trans. Reliab*. 1998, 47(2), 165-72

Levitin, G.;Lisnianski, A.; Elmakis, D. Structure optimization of power system with different redundant elements. *Electr. Power Syst. Res*. 1997, 43, 19-27

Denis-Papin, M.; Malgrange, Y. (1970), *Exerciții de calcul boolean cu soluțiile lor*, Ed. Tehnică, Bucharest, Romania

Jula, N.; Cepisca ,C.; Lungu, M.; Racuciu, C.; Ursu, T.; Raducanu, D. Theoretical and practical aspects for study and optimization of the aircrafts' electro energetic systems, *WSEAS Transactions on Circuits and Systems*, 12, Vol. 7, 2008, pp.999-1008

Chern, C.S., Jan, R.H. Reliability optimization problems with multiple constraints. *IEEE Trans. Reliab*.,1986,R-35, 431-6

Lyn, M.R. (1996). *Handbook of software reliability engineering*, New York, McGraw-Hill
    Billinton, R; Allan, R.N. (1996). *Reliability evaluation of power systems*, 2nd ed., New
    York, Plenum Press
Hecht, H. (2004). *System Reliability and Failure Prevention*, Artech House, London

Part 2

Aircraft Inspection and Maintenance

5

The Analysis of the Maintenance Process of the Military Aircraft

Mariusz Wazny
Military University of Technology
Poland

1. Introduction

This chapter presents the analysis of the maintenance process of a military aircraft with a detailed description of two areas, i.e. the process of maintaining and the process of operating. Each of these processes is briefly characterized. The section also involves methods enabling the determination of: residual durability of specified devices/systems of a military aircraft on the basis of the diagnostic parameters of these devices/systems, and the effectiveness of a combat task execution on the basis of information registered in the process of aiming. Each presented method is illustrated by a computational example.

2. Tasks executed by the military aircraft

A modern military aircraft (MMA) is a hybrid of the most up-to-date achievements in the field of materials engineering (the use of light metal alloys and composite structures), electronic engineering (fast microprocessor systems, modern systems in the field of power electronics), and specialized software supporting the maintenance process (automatic flight control system, integrated diagnostic systems). Due to such combination, tasks executed by MMA comprise a wide range that can be divided into two groups: with the use of aerial combat means and without the use of aerial combat means.

Depending on the nature of a mission, tasks including the use of aerial combat means can be generally classified as:

1. The gaining and maintenance of domination of airspace. This type of task is executed by fast and manoeuvrable aircrafts that are equipped with the most modern armament for aerial combat, i.e. air-to-air missiles and aircraft guns.
2. The support for the operations of ground forces and the navy. As regards this task, aircrafts equipped with air-to-ground weaponry, including rockets, bombs, and aircraft guns, play an important role.
3. The combating of a selected target of an air attack using precision-guided munitions launched from manned and unmanned aircrafts.

When analyzing the use of MMA in respect of the combat task realization without the use of aerial combat means, we can distinguish the following main tasks:

1. Air reconnaissance performed using both aircrafts equipped with specialized apparatus and unmanned flying objects configured for the performance of this type of a mission.
2. Air transport ensuring fast and efficient transfer of both infrastructure elements and soldiers into the area of a new localization for troops.

The support for the operations of different types of forces by means of, among other things, managing a mission on the basis of spatial information obtained via reconnaissance systems installed, for example, on an AWACS-type platform, or enabling the in-flight refuelling.

The analysis of the operations of the armed forces in recent armed conflicts indicates that MMAs are the basic element of the system of military operations. MMAs are used in the first instance to execute all of the above-mentioned tasks.

3. The organization of the maintenance process of the military aircraft

The technical objects maintenance is defined as a set of intentional organizational and economical operations of the people on the technical objects and the relationships between them from the beginning of the object lifecycle up to the end of lifecycle and object disposal. Relationships recognition and identification of the operations which appear between subjects based on the knowledge and experience of the technical objects designers, developers and engineers. The maintenance compliance and utility of product mainly depends on the engineers and designers crew professional competence. However the design presumptions can be altered many times during object lifecycle. These operations are performed to decrease maintenance "waste effect" and maximize "utility effect".

The modern military aircraft, which is the basic technical object in Polish Air Force organization structure, is the complex product including various constructional, technological, engineering and organizational concepts. Design of so sophisticated product based on tactical and technical military requirements which was created after modern battlefield analysis.

The aircraft construction is based on the module structure (Fig. 1) which allows dividing the specified tasks between separate functional blocks. This solution improves the maintenance process and facilitates service and operational use of the aircraft.

The conditions in which the aircrafts are operated are so specific that involves the specified requirements regarding high level of reliability, durability, effectiveness and safety parameters as far as airborne technology is concerned. Required levels of parameters are provided by determining specified functional structure of devices and specified level of redundancy.

Due to specific character of aircraft operations the aircraft maintenance can be performed only within specified system which provides the conditions indispensable for correct aircraft operation. This specified system is called Air System (AR) and contains the aircraft frame, the people who participate in the maintenance process and the devices building the system which ensure process permanence (in functional way) - Fig. 1.

The primary target in military aircraft maintenance process during peace is maintaining both the technical equipment and the personnel on the specified reliability and training level. It is required to provide high level of efficacy and effectiveness during wartime.

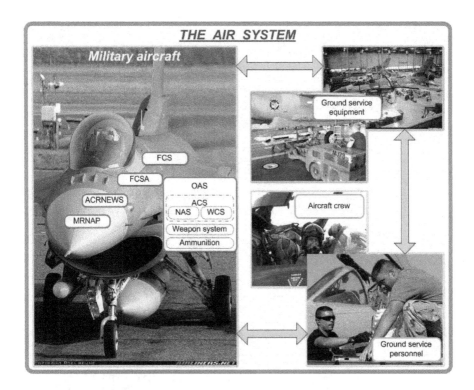

Fig. 1. Structural diagram of the military aircraft and the air system: FCSA – Flight Control System Actuators (frame construction with plating); FCS – Flight Control System; ACRNEWS – Airborne Communication, Radio Navigation and Electronic Warfare Systems.; MRNAP – Multifunctional Radar and Navigation and Aiming Pod; OAS – On-board Armament System; ACS – Armament Control System; WCS – Weapon Control System; NAS – Navigation and Aiming System.

Due to many various external factors, which influence negatively on the specified technical elements of the Air System, it can be claimed, that during operating process the elements are getting "used up". Therefore, due to maintain Air System in the appropriate reliability condition there is required to perform technical service. This action contains adjustment, tuning and replacement of particular devices or whole aggregates, in order to slow down the "using up" process.

In practice there are three aircraft maintenance strategies (Fig. 2.):

1. maintenance system containing prevention services schedule (recurring maintenance).
2. operational maintenance system.
3. preventive/predictive maintenance system.

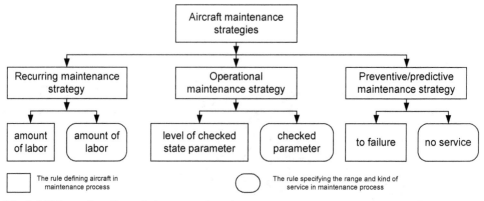

Fig. 2. Military aircrafts maintenance strategies.

Organization and scheme of military aircrafts recurring maintenance strategy is presented on Fig. 3. The basis of this maintenance strategy is the measurement of the amount of labor executed by the plant. As far as aircraft is concerned the amount of labor is defined as a number of hours in the sky.

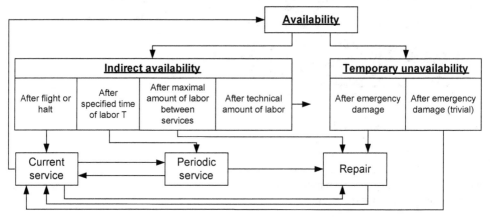

Fig. 3. Recurring maintenance strategy scheme.

One of the maintenance states in the recurring maintenance process is the indirect airworthiness state. The aircraft in this state is mostly working correctly but it lost the flying ability in order to circumstances determined on figure 3. After execution the specified amount of labor (hours of fly) the aircraft lifecycle should be either terminated or directed to the professional service to determine the new amount of labor possible to execute.

As far as operational maintenance strategy is concerned there is the rule the aircraft is in operation as long as the levels of specified parameters do not exceed the specified limits of error. The knowledge about the maintenance state of the device is determining by the external and internal diagnostic equipment. The service operations during this maintenance strategy are executed according to levels of measured diagnostic parameters. The proper control of operational maintenance strategy even for the considerable fleet of aircrafts requires control of every aircraft separately.

The preventive/predictive maintenance strategy defines the reliability as a designed characteristic. The level (value) of the reliability must be provided in the device design and manufacturing process and is maintaining during the device lifecycle. The maintenance schedule which is based on preventive/predictive maintenance strategy provides the desirable or defined levels of both reliability and flight safety. The all of described aircrafts maintenance strategies are followed during the real conditions fleet maintenance process.

Due to the development of diagnostic systems, military aircraft on-board systems include diagnostic procedures enabling the assessment of a current technical state of a given system. The procedure of assessing a given system is performed before an air operation. The procedure results provide information on a technical state of a military aircraft. Based on this information, a pilot decides either to perform a task or to withdraw from performing the task.

Apart from integrated diagnostic systems installed on board, there is a number of devices whose technical state is examined via monitoring and measuring equipment after its disassembly from the board of MMA. During maintenance works, diagnostic parameters of the examined devices are recorded and compared with the range of permissible changes. Any deviation beyond the assumed tolerance limits leads to the implementation of either appropriate maintenance procedures aiming at reducing the resultant deviation or appropriate corrections eliminating the deviation. The ability to predict the service life of MMA when diagnostic parameter tolerance might be exceeded would enable the appropriate management of the maintenance system of MMA. Thus, it is possible to optimize the time when MMA is under certain maintenance works and is not combat ready.

4. The process of maintaining the military aircraft

4.1 The influence of destructive factors on the technical state of devices used on the military aircraft

During the operation process of a military aircraft we can observe the change of technical parameters of selected devices along with the time of their operation. This change causes the deterioration of working conditions of a system and the loss of rated values of technical parameters. Factors influencing the above-mentioned changes include:

- changes of temperature and air-pressure,
- g-forces,
- vibrations,
- ageing process, etc.

The construction of technical systems is based on the assumption that a device fulfils its role when its operational/diagnostic parameters are within acceptable error limits. This assumption depends on the accuracy of work of particular system elements. Thus, in order to assure a faultless functioning of a military aircraft, we cannot allow operational parameters to exceed the acceptable error limits, which can be done in two ways: by frequent checks of operational parameter values of a device/system and its switch off when parameters are close to the fixed limit, or by determining the time after which operational parameters exceed values of the acceptable error.

The first way is onerous with regard to its organization and it is also time consuming and money consuming. Besides, the time spent on checking excludes a military aircraft from its use in a combat task, which consequently leads to a temporal decrease of the fighting efficiency of the air forces.

The second way is based on the use of a particular mathematical method enabling the description of value changes of operational parameters of a device/system and the evaluation of time in which a device/system is in operational state.

It is stated above that military aircrafts undergo changes during the exploitation of operational parameter values of particular devices in avionics system. The changes cause that operational parameter values approximate to the fixed acceptable limit. When parameter values equate with the limit value or exceed it, an adjustment must be done in order to restore nominal conditions of a device/system operation or the operation must be stopped. Figure 4 presents a theoretical course of changes of diagnostic parameter values.

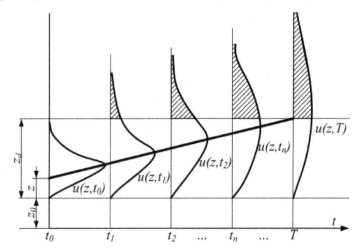

Fig. 4. Diagram of changes of diagnostic parameter values: z_0 – nominal value of a parameter, z – current value of a parameter, z_d – the limit of acceptable changes of parameter values

The second way is based on the use of a particular mathematical method enabling the description of value changes of operational parameters of a device/system and the evaluation of time in which a device/system is in operational state.

4.2 The model of diagnostic parameter changes in the aspect of the occurrence of destructive factors

In the figure, current value of a parameter is marked as "z". If $z < z_d$ then an element is fit for use, but if $z \geq z_d$ the elements losses its operational state. The change of diagnostic parameter values will be of a random character because of a specific character of MA operation process and the influence of destructive processes. So, let's consider "the wear of a device" of avionics system as a random process occurring during the operation of an aircraft.

Getting down to the analytical description of the diagram in Figure 4 and the determination of the density function of the changes of a diagnostic parameter values, the following assumptions were accepted:

1. The technical condition of an element is described by one diagnostic parameter which is marked as „z".
2. The change of the value of the parameter „z" happens only during the operation of a device, i.e. during the flight of an aircraft.
3. The parameter „z" is non-decreasing.
4. The change of the diagnostic parameter „z" is described by the following equation (1).

$$\frac{dz}{dN} = c \tag{1}$$

where:

c - random variable which depends on operational conditions of an element;
N - the number of flights of an aircraft.

1. If $z \in [0, z_d]$ then an element is fit for use, in other case the element is considered as unfit for use.
2. The intensity of flights of an aircraft is described by the following dependence (2).

$$\lambda = \frac{P}{\Delta t} \tag{2}$$

where:

Δt - the range of time in which the flight of an aircraft can be performed with the probability P,
P - the probability of the flight performance within the time interval with length Δt.

The time interval with length Δt shall be selected in such a way as to fulfil the following inequality (3).

$$\lambda \Delta t \leq 1 \tag{3}$$

The intensity of flights λ enables the determination of the number of flights of an aircraft up to the moment t form the following formula:

$$N = \lambda t \tag{4}$$

Using the formula (4), the equation (1) can be written in the following form:

$$\frac{dz}{dt} = \lambda c \tag{5}$$

The dynamics of the changes of a diagnostic parameter can be described by the following difference equation (6).

$$U_{z,t+\Delta t} = (1 - \lambda\Delta t)U_{z,t} + \lambda\Delta t\, U_{z-\Delta z,t} \tag{6}$$

where:

$U_{z,t}$ - the probability that in the moment t the value of a diagnostic parameter will be z;

Δz - the increment of the diagnostic parameter z during one flight of an aircraft.

The functional notation of the equation (6) has the following form:

$$u(z,\ t+\Delta t) = (1 - \lambda\Delta t)\, u(z,t) + \lambda\Delta t\, u(z - \Delta z, t) \tag{7}$$

where:

$u(z,t)$ - the density function of the probability of the diagnostic parameter value z in the moment t;

$(1 - \lambda\Delta t)$ - the probability that in the time interval with length Δt the flight will not be performed;

$\lambda\Delta t$ - the probability of the flight performance in the time interval with length Δt.

The equation (7) was transformed by substituting the following differential equation (8).

$$\frac{\partial u(z,\ t)}{\partial z} = -\lambda\Delta z\, \frac{\partial u(z,\ t)}{\partial z} + \frac{1}{2}\lambda(\Delta z)^2\, \frac{\partial^2 u(z,\ t)}{\partial z^2} \tag{8}$$

where: $\Delta z = c$.

Due to the fact that c is a random variable, the following mean value was introduced:

$$E[c] = \int\limits_{c_d}^{c_g} c\, f(c)\, dc \tag{9}$$

where: $f(c)$ - the density function of the random variable c;

c_g, c_u - the limits of variation of c.

Taking into consideration the dependence (9), the differential equation (8) can be written in the following form:

$$\frac{\partial u(z,\ t)}{\partial t}\Delta t = -\lambda E[c]\frac{\partial u(z,\ t)}{\partial z} + \frac{1}{2}\lambda(E[c])^2\, \frac{\partial^2 u(z,\ t)}{\partial z^2} \tag{10}$$

where: $\lambda E[c]$ - the mean increment of the parameter value per time unit;

$\lambda(E[c])^2$ - the mean square increment of the value of the diagnostic parameter per time unit.

The solution of the equation (10) is the unknown density function of the probability of the random variable z in the following form:

$$u(z,t) = \frac{1}{\sqrt{2\pi A(t)}}\, e^{-\frac{(z-B(t))^2}{2A(t)}} \tag{11}$$

where:

$$B(t) = \int_0^t \lambda E[c]\,dt = \lambda E[c]t, \quad A(t) = \int_0^t \lambda \left(E[c]\right)^2 dt = \lambda E[c]^2\, t \tag{12}$$

Assuming that:

$$b = \lambda E[c], \quad a = \lambda E[c]^2 \tag{13}$$

the density function (11) has the following form:

$$u(z,t) = \frac{1}{\sqrt{2\pi a t}}\, e^{-\frac{(z-bt)^2}{2at}} \tag{14}$$

The dependence (14) is the probabilistic characterisation of the increase of the wear in the function of the flying time. However, it is important to know the distribution of the time (the flying time) of the exceedance of the acceptable error value of the parameter z.

The probability of the exceedance of the acceptable value by the current value of the diagnostic parameter „z" can be written in the following form:

$$Q(t;z_d) = \int_{z_d}^{\infty} \frac{1}{\sqrt{2\pi a t}}\, e^{-\frac{(z-bt)^2}{2at}}\, dz \tag{15}$$

The density function of the time distribution of the exceedance of the acceptable state z_d has the following form:

$$f(t) = \frac{\partial}{\partial t}\, Q(t;z_d) \tag{16}$$

Thus

$$f(t) = \frac{\partial}{\partial t} \int_{z_d}^{\infty} \frac{1}{\sqrt{2\pi a t}}\, e^{-\frac{(z-bt)^2}{2at}}\, dz \tag{17}$$

$$f(t) = \int_{z_d}^{\infty} \left\{ \frac{\partial}{\partial t} \left[\frac{1}{\sqrt{2\pi a t}}\, e^{-\frac{(z-bt)^2}{2at}} \right] \right\} dz \tag{18}$$

After calculating the derivative, we obtain:

$$f(t)_{z_d} = \int_{z_d}^{\infty} \left[u(z,t) \left(\frac{z^2 - b^2 t^2 - at}{2at^2} \right) \right] dz \tag{19}$$

The original function with regard to the integrand of the dependence (19) has the following form (20).

$$w(z,t) = u(z,t) \left(-\frac{z+bt}{2t} \right) \tag{20}$$

We calculate the integral (19).

$$f(t)_{z_d} = u(z,t) \left(-\frac{z+bt}{2t} \right) \Bigg|_{z_d}^{\infty} = \frac{z_d + bt}{2t} \frac{1}{\sqrt{2\pi at}} e^{-\frac{(z_d - bt)^2}{2at}} \tag{21}$$

Thus, the dependence (21) determines the density function of the time of the first transition of the current value of the parameter „z" through the acceptable state.

Having the above-mentioned data, we can determine the durability of a device with respect to the change of the value of the parameter z. For this purpose, we can write down that the formula for the reliability of a device has the following form:

$$R(t) = 1 - \int_0^t f(t)_{z_d} dt \tag{22}$$

where the density function $f(t)_{z_d}$ is determined by the formula (21).

The unreliability of a device can be determined from the dependence (23).

$$Q(t) = \int_0^t \frac{z_d + bt}{2t} \cdot \frac{1}{\sqrt{2\pi at}} e^{-\frac{(z_d - bt)^2}{2at}} dt \tag{23}$$

The integral (23) has to be simplified. It can be observed that the integrand can be written in the following form:

$$\frac{z_d + bt}{2t} \cdot \frac{1}{\sqrt{2\pi at}} e^{-\frac{(z_d - bt)^2}{2at}} = \frac{z_d + bt}{2t} \cdot \frac{1}{\sqrt{2\pi at}} e^{-\frac{(bt - z_d)^2}{2at}} \tag{24}$$

and now we have to solve the indefinite integral.

$$\int \frac{(z_d + bt)}{2t} \cdot \frac{1}{\sqrt{2\pi at}} e^{-\frac{(bt - z_d)^2}{2at}} dt \tag{25}$$

We make the substitution in the above-mentioned integral.

$$\frac{(bt - z_d)^2}{2at} = u \tag{26}$$

Thus

$$\frac{du}{dt} = \frac{bt + z_d}{2at^2}(bt - z_d) \tag{27}$$

$$dt = \frac{2at^2}{(bt + z_d)(bt - z_d)} du \tag{28}$$

After the substitution, the integral (25) has the following form (29).

$$\int \frac{z_d + bt}{2t} \cdot \frac{1}{\sqrt{2\pi at}} e^{-u} \cdot \frac{2at^2}{(bt + z_d)(bt - z_d)} du = \frac{1}{2\sqrt{\pi}} \int \frac{1}{\sqrt{u}} e^{-u} du \tag{29}$$

Then, we make the second substitution.

$$\sqrt{u} = w, \rightarrow \frac{dw}{du} = \frac{1}{2\sqrt{u}}, \rightarrow \frac{du}{dw} = 2w, \rightarrow du = 2w\,dw \tag{30}$$

Taking into consideration the above-mentioned dependencies, the integral (29) can be written in the following form:

$$\frac{1}{2\sqrt{\pi}} \int \frac{1}{w} e^{-w^2} 2w\,dw = \frac{1}{\sqrt{\pi}} \int e^{-w^2} dw \tag{31}$$

We make one more substitution.

$$w^2 = \frac{y^2}{2}, \rightarrow 2w\,dw = y\,dy, \rightarrow dw = \frac{y}{2w}dy, \rightarrow dw = \frac{y}{\sqrt{2}} \tag{32}$$

Thus, we obtain the integral in the following form:

$$\frac{1}{\sqrt{2\pi}} \int e^{-\frac{y^2}{2}} dy \tag{33}$$

where:

$$y = \frac{bt - z_d}{\sqrt{at}} \tag{34}$$

Substituting the results into the formula (22) and remembering the appropriate notation of the integration limits, we obtain the formula for the reliability:

$$R(t) = 1 - \frac{1}{\sqrt{2\pi}} \int_{-\infty}^{\frac{bt-z_d}{\sqrt{at}}} e^{-\frac{y^2}{2}} \, dy \qquad (35)$$

The distribution function for the standard normal distribution has the following form (36).

$$\Phi(x) = \frac{1}{\sqrt{2\pi}} \int_{-\infty}^{x} e^{-\frac{y^2}{2}} \, dy \qquad (36)$$

Finally, the formula for the reliability of a system has the form of the following dependence:

$$R^*(t) = 1 - \Phi\left(\frac{b^*t - z_d}{\sqrt{a^*t}}\right) \qquad (37)$$

where b^* and a^* are coefficients after the estimation on the basis of data obtained from the exploitation of military aircrafts.

Thus, the risk of a device damage can be determined from the following dependence (38).

$$Q^* = 1 - R^*(t) = \Phi(\gamma) \qquad (38)$$

where:

$$\gamma = \frac{b^*t - z_d}{\sqrt{a^*t}} \qquad (39)$$

Assuming a specified level of damage risk, we can find γ (by reading values on the tables of the normal distribution). Knowing the value of γ, we can determine the durability (i.e. t) from the dependence (39). For this purpose, the dependence (39) was transformed into the following square equation (40).

$$b^{*2}t^2 - \left(\gamma^2 a^* + 2b^* z_d\right)t + z_d^2 = 0 \qquad (40)$$

Thus, the durability:

$$T = \frac{\left(\gamma^2 a^* + 2b^* z_d\right) - \sqrt{\left(2b^* z_d + \gamma^* a^*\right)^2 - 4b^{*2} z_d^2}}{2b^{*2}} \qquad (41)$$

4.3 A computational example

The efficiency of the chosen system is determined with the help of diagnostic parameters describing the technical condition of particular devices of the system. An aiming head (a navigation and aiming device) is an important device of avionics system. Its technical condition is described by two diagnostic parameters: ε and β which describe the coordinates of position of sight marker.

On the basis of analyzing results of checks of a particular population of aiming heads it was established that as the time of operation goes by and as a result of the influence of destructive factors, the values of these parameters undergo changes. Table 1 presents an exemplary course of changes of values of the diagnostic parameters ε and β during an operation process.

T [months]	0	27	40	57	83	94	102	110	116
ε	0	0,01	0,01	0,01	0,07	0,48	0,48	0,54	0,73
β	0	0,23	0,26	0,26	0,39	0,50	0,53	0,56	0,59

Table 1. Changes of diagnostic parameter values in an aiming head during an operation process

Having data describing the values of deviation of a diagnostic parameter in the following form $[(z_0,t_0),(z_1,t_1),(z_2,t_2),...,(z_n,t_n)]$, and basing on the following formulas,

$$b^* = \frac{z_n}{t_n}, \quad a^* = \frac{1}{n}\sum_{k=0}^{n-1}\frac{\left[(z_{k+1}-z_k)-b^*(t_{k+1}-t_k)\right]^2}{(t_{k+1}-t_k)} \tag{42}$$

the values of the density function coefficients for both diagnostic parameters were determined:

$$a_\varepsilon^* = 0,002; \quad b_\varepsilon^* = 0,0063; \quad a_\beta^* = 0,0003; \quad b_\beta^* = 0,0051 \tag{43}$$

Assuming the following level of reliability $R^*(t) = 0,99$, the value of the parameter $\gamma = 2,32$ was read on the tables of normal distribution. The parameter z_d was determined on the basis of a technical documentation which is used for service works and includes information on the acceptable values of deviations of the diagnostic parameters.

The values of the parameters a, b, γ, z_d were substituted into the equation (41), and the time after which the values of the diagnostic parameter deviations exceed the limit state was calculated. In this case, the time comes to:

$$T_\varepsilon = 5[\text{months}], \quad T_\beta = 33[\text{months}] \tag{44}$$

since the last check of the diagnostic parameters. The values (44) can be used in technical service depending on the adopted service strategy.

Summing up, we can state that the above-presented method seems to be correct and enables the analysis of a device/system technical condition with respect to the character of changes of values of the diagnostic parameters. The above-presented calculation example enabled the verification of the developed model and showed application qualities of the method. This method can be useful in future work on the improvement of both the operation process and the way of use of aircrafts with avionics system because it enables the determination of time during which a device is fit for use.

Moreover, due to its universal character, the method can be used to determine the residual life of any technical object whose technical condition is determined by analyzing values of the diagnostic parameters.

5. The process of operating the military aircraft

5.1 The influence of destructive factors on the course of the process of operating the military aircraft

The use of military aircrafts concerns mainly the performance of a particular combat task, which often involves the use of aerial combat means. As far as an airborne function of a military aircraft is concerned, the main stages of its operation comprise the take-off, the staying in the air, and the landing. On the other hand, when analyzing the process of the operation of the on-board armament system, we can assume that the operational effect is the sum of the partial effects gained during the flight phase in relation to:

- target detection;
- the execution of the aiming process;
- the execution of the process of attacking.

The level of effect of munitions on a target is the most commonly assumed rate that characterizes the operational effect obtained during the execution of a combat task involving the use of aerial combat means. As regards the on-board armament system, the obtained effect comes down to the determination of the difference between the value of target coordinates and the coordinate values of a drop point of combat armament.

Based on the structural diagram (Fig. 1) and the functions of the on-board armament system, we can assume that the Armament Control System (ACS) is the basic element that affects the value of the operational effect. Both at the stage of maintenance and operation, ACS provides information that is essential for the accurate functioning of the on-board armament system (OAS). In turn, as regards the ACS, its most crucial element involves the navigation and aiming system (NAS). Its basic task comprises the realization of a set of algorithms. Their solution enables – in the maintenance system - the reconstruction of the nominal values of particular initial parameters; - in the operation system – the proper usage of combat means (the intended use). The latter system is the subject of further discussion.

The analysis of the operational effect can be performed on the basis of the assessment of conditions in which NAS is used and the determination of causes that have a negative impact on the final value of the obtained effect. As regards NAS, during the execution of a combat task, the operational effect is the total angular correction represented as an aiming indicator in a pilot's field of view. The process of aiming and attacking is executed on the basis of the total angular correction. Thus, we can assume that the assessment of the operational effect involves the determination of accuracy in defining and reproducing the position of a moving aiming indicator.

The next aspect concerns the use of the aiming correction by a pilot. When the correction is defined and illustrated, the task comes down to the determination of the flight conditions in which an aiming indicator coincides with a target at the moment of using combat means. Based on the conducted analysis, we can assume that the execution of a combat task under real conditions is not an easy process. The causes of errors affecting the value of the operational effect connected with the aiming process execution can be represented as the equation for the pooled error of the aiming process execution Δ_Σ:

$$\Delta_\Sigma =(\Delta_M +\Delta_K +\Delta_I +\Delta_A)+ (\Delta_C +\Delta_W +\Delta_R +\Delta_O) +\Delta_N \tag{45}$$

The error of the method for solving the aiming-related equations Δ_M characterizes two groups of causes:

1. connected with the relative uncertainty resulting from the processing of initial data concerning the aiming process by NAS functional elements, and
2. concerning the error function of equations for aiming.

The system configuration error Δ_K connects with entering invalid control signals (that characterize the combat task being performed) into NAS.

The instrumental error Δ_I connects with the accuracy of determining the operational parameters of NAS by particular information transmitters. This error concerns mainly the measurement error.

The reconstruction error Δ_A characterizes the adequacy of a physical combat situation taking place during the execution of the aiming process to the assumed attack diagram which was used to determine the aiming equations.

The causes of variance between the aiming indicator position and the target Δ_C result from an incorrect approach of an aircraft to an attack path.

The causes of the failure to maintain the required conditions for aiming and attacking Δ_W connect with the failure to keep the required angle of diving, flight speed, bank angle, etc., i.e. the exceeding of the nominal values of particular parameters describing a combat task.

The effect of the weapon position Δ_R on the pooled error value Δ_Σ, concerns mainly the process of aiming during the execution of the process of attacking with the use of aerial combat means (that are applied in a time series of particular length).

Environmental conditions determining the value of the error Δ_O significantly influence the execution of the aiming process. Due to the fact that an aircraft moves at high speed in a heterogeneous space, it may encounter various conditions prevailing in space layers or areas, which directly translates into the perturbation of flight-related parameter values.

The general error Δ_N concerns causes which are not included in the presented classification and are the resultant of the lack of possibility to learn or describe them in an analytical way at the present state of knowledge.

All the above-mentioned errors can be of two kinds: determined errors (systematic errors) and probabilistic errors (random errors). So, their accumulated form Δ_Σ will be burdened with both types of errors. The phenomenon of the random error occurrence is not precisely determined, that is why an attempt to evaluate its value is fully justified. A random character of compound errors causes that the operational effect of MMA application is burdened with the random error, too.

5.2 The model of the assessment of the execution of a combat mission by the military aircraft

The execution of the aiming process generally comes down to the process of making an aiming indicator coincide with a target. Significant elements of this process include parameters that determine the aiming indicator position and a set of actions aiming at pointing the indicator at a target. Based on these elements, we can consider the process of aiming as the execution of the process of building the aiming triangle using: a pilot – the system operator, an aiming indicator – the quantity describing the appropriate spatial orientation of an aircraft, and a target – the basic point in the execution of the aiming process. The aim of the process is to align these three elements.

The aiming correction is obtained by recording particular parameters (necessary to solve aiming equations) and processing them in NAS. The aiming correction value is represented as the central point of a moving aiming indicator which is displayed on the reflector of the sight head. Due to the effect of various constraints, the aiming indicator can adopt different positions in the assumed flat coordinate system (Fig. 6) placed on the plane of the sight head reflector. The indicator can either move in one out of four directions or move back to the previously occupied position.

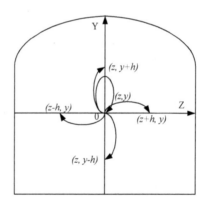

Fig. 6. A graphical representation of the occurrence of possible deviations of the central point of the moving indicator during the execution of the aiming process

$U_{z,y,t}$ denotes the probability that at the moment t the position deviations of the central point of the moving indicator are z and y, where t is the current time of the process of aiming. This probability is characterized by the density function denoted as $U(z,y,t)$. Therefore, using the density function $U(z,y,t)$ we can describe the dynamics of changes in the position deviations of the central point of the moving indicator by a difference equation.

Regarding the issue being discussed above, the difference equation is as follows:

$$U(z,y,t+\Delta t) = P_{00}U(z,y,t) + P_{10}U(z-h,y,t) + P_{20}U(z+h,y,t) + \\ + P_{01}U(z,y-h,t) + P_{02}U(z,y+h,t)$$

(46)

where:

$U(z,y,t)$ - the probability density function of deviation values at the moment t;
Δt - the time value between the specified deviations;
h - the deviation value along the specified axes;
P_{00} - the probability that the deviation value will not change;
P_{10} - the probability that the deviation value along the OZ axis will change by $-h$ at the time Δt;
P_{20} - the probability that the deviation value along the OZ axis will change by h at the time Δt;
P_{01} - the probability that the deviation value along the OY axis will change by $-h$ at the time Δt;
P_{02} - the probability that the deviation value along the OY axis will change by h at the time Δt;

When we use expressions obtained from the expansion of the function $U(z,y,t)$ in the Taylor series in the surrounding of the point (z,y) and the time t in accordance with the relationships of the following set of equations:

$$\left.\begin{aligned}
U(z,y,t+\Delta t) &= U + \frac{\partial U}{\partial t}\Delta t \\[6pt]
U(z-h,y,t) &= U - \frac{\partial U}{\partial z}h + \frac{1}{2}h^2\frac{\partial^2 U}{\partial z^2} \\[6pt]
U(z+h,y,t) &= U + \frac{\partial U}{\partial z}h + \frac{1}{2}h^2\frac{\partial^2 U}{\partial z^2} \\[6pt]
U(z,y-h,t) &= U - \frac{\partial U}{\partial y}h + \frac{1}{2}h^2\frac{\partial^2 U}{\partial y^2} \\[6pt]
U(z,y+h,t) &= U + \frac{\partial U}{\partial y}h + \frac{1}{2}h^2\frac{\partial^2 U}{\partial y^2}
\end{aligned}\right\} \tag{47}$$

where $U=U(z,y,t)$, and the fact that $P_{00}+P_{10}+P_{20}+P_{01}+P_{02}=1$, the equation (47) takes the following form:

$$U + \frac{\partial U}{\partial t}\Delta t = P_{00}U + P_{10}\left(U - \frac{\partial U}{\partial z}h + \frac{1}{2}h^2\frac{\partial^2 U}{\partial z^2}\right) + P_{20}\left(U + \frac{\partial U}{\partial z}h + \frac{1}{2}h^2\frac{\partial^2 U}{\partial z^2}\right) +$$
$$+ P_{01}\left(U - \frac{\partial U}{\partial y}h + \frac{1}{2}h^2\frac{\partial^2 U}{\partial y^2}\right) + P_{02}\left(U + \frac{\partial U}{\partial y}h + \frac{1}{2}h^2\frac{\partial^2 U}{\partial y^2}\right) \tag{48}$$

When adding and subtracting U in the equation (48) and multiplying appropriate expressions in the brackets and taking the parameter U outside the brackets, the following result was obtained:

$$\frac{\partial U}{\partial t}\Delta t = -U + (P_{00}+P_{10}+P_{20}+P_{01}+P_{02})U + P_{10}\left(-\frac{\partial U}{\partial z}h + \frac{1}{2}h^2\frac{\partial^2 U}{\partial z^2}\right) +$$
$$+ P_{20}\left(\frac{\partial U}{\partial z}h + \frac{1}{2}h^2\frac{\partial^2 U}{\partial z^2}\right) + P_{01}\left(-\frac{\partial U}{\partial y}h + \frac{1}{2}h^2\frac{\partial^2 U}{\partial y^2}\right) + \tag{49}$$
$$+ P_{02}\left(\frac{\partial U}{\partial y}h + \frac{1}{2}h^2\frac{\partial^2 U}{\partial y^2}\right)$$

Using the assumption that the sum of all probabilities describing the weapon angular position equals one, the equation (49) takes the following form:

$$\frac{\partial U}{\partial t}\Delta t = -P_{10}\frac{\partial U}{\partial z}h + P_{10}\frac{1}{2}h^2\frac{\partial^2 U}{\partial z^2} + P_{20}\frac{\partial U}{\partial z}h + P_{20}\frac{1}{2}h^2\frac{\partial^2 U}{\partial z^2} - P_{01}\frac{\partial U}{\partial y}h +$$
$$+ P_{01}\frac{1}{2}h^2\frac{\partial^2 U}{\partial y^2} + P_{02}\frac{\partial U}{\partial y}h + P_{02}\frac{1}{2}h^2\frac{\partial^2 U}{\partial y^2}$$

(50)

After grouping the quantities from the above equation, the following equation was obtained:

$$\frac{\partial U}{\partial t}\Delta t = -P_{10}\frac{\partial U}{\partial z}h + P_{20}\frac{\partial U}{\partial z}h + P_{10}\frac{1}{2}h^2\frac{\partial^2 U}{\partial z^2} + P_{20}\frac{1}{2}h^2\frac{\partial^2 U}{\partial z^2} - P_{01}\frac{\partial U}{\partial y}h +$$
$$+ P_{02}\frac{\partial U}{\partial y}h + P_{01}\frac{1}{2}h^2\frac{\partial^2 U}{\partial y^2} + P_{02}\frac{1}{2}h^2\frac{\partial^2 U}{\partial y^2}$$

(51)

After dividing both sides of the equation (51) by Δt, the following result was obtained:

$$\frac{\partial U}{\partial t} = -\frac{(P_{10} - P_{20})h}{\Delta t}\frac{\partial U}{\partial z} + \frac{(P_{10} + P_{20})\frac{1}{2}h^2}{\Delta t}\frac{\partial^2 U}{\partial z^2} +$$
$$-\frac{(P_{01} - P_{02})h}{\Delta t}\frac{\partial U}{\partial y} + \frac{(P_{01} + P_{02})\frac{1}{2}h^2}{\Delta t}\frac{\partial^2 U}{\partial y^2}$$

(52)

By introducing the following denotations:

$$b_1 = \frac{(P_{10} - P_{20})h}{\Delta t}, \quad b_2 = \frac{(P_{01} - P_{02})h}{\Delta t}$$

(53)

$$a_1 = \frac{(P_{10} + P_{20})h^2}{\Delta t}, \quad a_2 = \frac{(P_{01} + P_{02})h^2}{\Delta t}$$

(54)

and substituting them into the equation (52), the following differential equation was obtained:

$$\frac{\partial U}{\partial t} = -b_1\frac{\partial U}{\partial z} - b_2\frac{\partial U}{\partial y} + \frac{1}{2}a_1\frac{\partial^2 U}{\partial z^2} + \frac{1}{2}a_2\frac{\partial^2 U}{\partial y^2}$$

(55)

The following function is the solution of the above equation:

$$U(z,y,t) = \frac{1}{\sqrt{2\pi a_1 t}\sqrt{2\pi a_2 t}}e^{-\frac{1}{2}\left(\frac{(z-b_1 t)^2}{a_1 t} + \frac{(y-b_2 t)^2}{a_2 t}\right)}$$

(56)

Assuming that the probabilities P_{10} and P_{20} are of the same order, i.e. $P_{10}=P_{20}$, we can write that the coefficient $b_1\approx0$. Similarly, we can assume that the probabilities P_{01} and P_{02} are also of the same order, so the coefficient $b_2\approx0$. Given these assumptions, the equation (55) takes the following form:

$$\frac{\partial U}{\partial t} = \frac{1}{2}a_1\frac{\partial^2 U}{\partial z^2} + \frac{1}{2}a_2\frac{\partial^2 U}{\partial y^2} \tag{57}$$

The following form of the density function is the solution of the equation (57):

$$U(z,y,t) = \frac{1}{\sqrt{2\pi a_1 t}\sqrt{2\pi a_2 t}}e^{-\frac{1}{2}\left(\frac{z^2}{a_1 t}+\frac{y^2}{a_2 t}\right)} \tag{58}$$

The explicit form of the density function (58) requires determining the equation coefficients (57) and connects with:

-   obtaining input data;
-   determining the density function (58);
-   determining the likelihood function L enabling the determination of the parameter estimates a_1 and a_2:

$$L = \frac{1}{(2\pi)^n (a_1 a_2)^{\frac{n}{2}}}\prod_{k=1}^{n-1}\frac{1}{(t_{k+1}-t_k)}\exp\left\{-\frac{1}{2}\left[\frac{(z_{k+1}-z_k)^2}{a_1(t_{k+1}-t_k)}+\frac{(y_{k+1}-y_k)^2}{a_2(t_{k+1}-t_k)}\right]\right\} \tag{59}$$

To determine the parameters a_1 and a_2 we can use the method of the maximum likelihood. The method consists in finding the parameter values a_1 and a_2 that maximize the likelihood function. So, we seek the solution of the set of equations

$$\begin{cases}\dfrac{\partial \ln L}{\partial a_1} = 0 \\[2mm] \dfrac{\partial \ln L}{\partial a_2} = 0\end{cases} \tag{60}$$

Therefore, the logarithm of the likelihood function L takes the following form:

$$\ln L = -n\ln 2\pi - \frac{n}{2}\ln a_1 - \frac{n}{2}\ln a_2 +$$
$$+\sum_{k=1}^{n-1}\left[\ln(t_{k+1}-t_k)+\left[-\frac{1}{2}\left(\frac{(z_{k+1}-z_k)^2}{a_1(t_{k+1}-t_k)}+\frac{(y_{k+1}-y_{*k})^2}{a_2(t_{k+1}-t_k)}\right)\right]\right] \tag{61}$$

By determining the derivatives of the function L relative to specified parameters, the following set of equations was obtained:

$$\begin{cases} -\dfrac{n}{2a_1} + \sum_{k=1}^{n-1} \dfrac{(z_{k+1} - z_k)^2}{2a_1^2(t_{k+1} - t_k)} = 0 \\[4mm] -\dfrac{n}{2a_2} + \sum_{k=1}^{n-1} \dfrac{(y_{k+1} - y_k)^2}{2a_2^2(t_{k+1} - t_k)} = 0 \end{cases} \tag{62}$$

which after transformation provides the following equations (63):

$$\begin{cases} a_1 = \dfrac{1}{n} \sum_{k=1}^{n-1} \dfrac{(z_{k+1} - z_k)^2}{(t_{k+1} - t_k)} \\[4mm] a_2 = \dfrac{1}{n} \sum_{k=1}^{n-1} \dfrac{(y_{k+1} - y_k)^2}{(t_{k+1} - t_k)} \end{cases} \tag{63}$$

Therefore, the parameters a_1 and a_2 can be defined on the basis of the above set of equations. When analyzing the function notation (58), it can be assumed that in order to determine the variance characterizing the distribution of the indicator central point, the parameters a_1 and a_2 must be multiplied by time, which leads to the following result:

$$\begin{cases} \sigma_z^2(t_n) = a_1 t_n = \dfrac{1}{n} \sum_{k=1}^{n-1} \dfrac{(z_{k+1} - z_k)^2}{(t_{k+1} - t_k)} \sum_{k=1}^{n-1}(t_{k+1} - t_k) \\[4mm] \sigma_y^2(t_n) = a_2 t_n = \dfrac{1}{n} \sum_{k=1}^{n-1} \dfrac{(y_{k+1} - y_k)^2}{(t_{k+1} - t_k)} \sum_{k=1}^{n-1}(t_{k+1} - t_k) \end{cases} \tag{64}$$

The determination of the function parameters (58) will allow defining the probability density function of the correct position of the indicator central point.

As regards the case described, it is assumed that the probability of the occurrence of deviations in any direction of the assumed coordinate axes is the same. Such situation takes place when the process of aiming is performed correctly, i.e. when at the beginning of the aiming process, an aiming indicator coincides with a target and any dislocation of the indicator is compensated with its resetting on the target. A real process of aiming often involves the indicator dislocation relative to a target. The occurrence of such dislocation causes that the probability of the indicator dislocation in a specified direction is higher than the indicator dislocation in an opposite direction. Thus, the values of the parameters b_1 and b_2 are not 0. Therefore, the differential equation describing the aiming process takes the form of the equation (55). Its solution is the density function (56). The parameters b_1, b_2, a_1 and a_2 need to be determined for the function. Using the above-described technique, the likelihood function (65) was determined. It was used to estimate the sought parameters:

$$L = \frac{1}{(2\pi)^n (a_1 a_2)^{\frac{n}{2}}} \prod_{k=1}^{n-1} \frac{1}{(t_{k+1} - t_k)} \exp\left\{ -\frac{1}{2} \left[\begin{array}{l} \dfrac{((z_{k+1} - z_k) - b_1(t_{k+1} - t_k))^2}{a_1(t_{k+1} - t_k)} + \\[4mm] \dfrac{((y_{k+1} - y_k) - b_2(t_{k+1} - t_k))^2}{a_2(t_{k+1} - t_k)} \end{array} \right] \right\} \tag{65}$$

The process of determining the function parameter (65) is analogous to the way of determining the equation coefficients (59). By determining the derivatives of the function logarithms (65) relative to specified coefficients and comparing them to 0, the following relationships were obtained:

$$b_1 = \frac{z_n}{t_n}, \qquad\qquad b_2 = \frac{y_n}{t_n}$$

$$a_1 = \frac{1}{n}\sum_{k=1}^{n-1} \frac{\left[(z_{k+1}-z_k)-b_1(t_{k+1}-t_k)\right]^2}{(t_{k+1}-t_k)} \qquad (66)$$

$$a_2 = \frac{1}{n}\sum_{k=1}^{n-1} \frac{\left[(y_{k+1}-y_k)-b_2(t_{k+1}-t_k)\right]^2}{(t_{k+1}-t_k)}$$

By determining the values of the above coefficients and substituting them into the equation (56), we can determine the density function of the indicator position during the aiming process involving the indicator dislocation relative to a target.

The indicator path relative to a target (described for subsequent moments t_0, t_1, t_2, ..., t_n,) can be characterized by horizontal coordinates z_0, z_1, z_2, ..., z_n and vertical coordinates y_0, y_1, y_2, ..., y_n of the assumed coordinate system. When converting these quantities to current data, the time of recording the position of the aiming indicator can be replaced by the number of the registered positions (next coordinate values will constitute the sum of previous coordinates). Thus, the indicator position will be characterized by:

1. the number of registered positions: 0, 1, 2, ..., n;

2. the deviation toward the 0Z axis: 0, z_1, (z_1+z_2), $(z_1+z_2+z_3)$, ..., $\sum_{i=1}^{n} z_i$;

3. the deviation toward the 0Y axis: 0, y_1, (y_1+y_2), $(y_1+y_2+y_3)$, ..., $\sum_{i=1}^{n} y_i$.

Based on the above, we can determine the following parameters:

$$b_1^* = \frac{\sum_{i=1}^{n} z_i}{n}, \qquad\qquad b_2^* = \frac{\sum_{i=1}^{n} y_i}{n} \qquad (67)$$

$$\sigma_1^2 = a_1^* = \frac{1}{n}\sum_{k=1}^{n-1}\left[(\hat{z}_{k+1}-\hat{z}_k)-\left(\frac{1}{n}\sum_{i=1}^{n} z_i\right)\right]^2$$

$$\sigma_2^2 = a_2^* = \frac{1}{n}\sum_{k=1}^{n-1}\left[(\hat{y}_{k+1}-\hat{y}_k)-\left(\frac{1}{n}\sum_{i=1}^{n} y_i\right)\right]^2 \qquad (68)$$

where:

$$\hat{z}_{k+1} = \sum_{i=1}^{k+1} z_i, \quad \hat{z}_k = \sum_{i=1}^{k} z_i$$

$$\hat{y}_{k+1} = \sum_{i=1}^{k+1} y_i, \quad \hat{y}_k = \sum_{i=1}^{k} y_i \qquad (69)$$

Because

$$\hat{z}_{k+1} - \hat{z}_k = z_{k+1} \quad \text{and} \quad \hat{y}_{k+1} - \hat{y}_k = y_{k+1} \qquad (70)$$

therefore:

$$\sigma_1^2 = \frac{1}{n}\sum_{k=1}^{n}\left[z_k - \frac{1}{n}\sum_{i=1}^{n} z_i \right]^2$$

$$\sigma_2^2 = \frac{1}{n}\sum_{k=1}^{n}\left[y_k - \frac{1}{n}\sum_{i=1}^{n} y_i \right]^2 \qquad (71)$$

The above relationships can be used to describe the process of aiming under real-life conditions.

5.3 A computational example

The execution of a combat task with the use of aerial combat means is characterized by the fact that the possibility of their use is determined by conditions that constitute a set of various factors enabling the performance of a combat task at the required level and with the consideration of a current tactical, navigational, meteorological, and radio-technical situation. The basic determinants of these conditions involve combat capabilities of an aircraft and the level of competence among aircrew members. The essence of the aiming process comes down to the controlling of an aircraft in such a way that it reaches the point in space where the applied weapon will hit a target. This procedure is performed in the NAS environment on the basis of the following data:

- motion parameters of an aircraft executing an attack, a target, and parameters of the centre where an aircraft motion is executed;
- the required coordinates of a target;
- the actual coordinates of a target;
- the comparison between actual and required coordinates of a target.

A common method for analyzing the aiming process during an attack is the recorded material analysis (using either the film placed in a photo-control apparatus located in front of the sight head or a camera recording a tactical situation in front of MMA.) Based on the recorded material, it is possible to determine a mutual position of an aiming indicator and a target at the moment of a weapon use.

Having the material registered by photo-control devices (Fig. 6) and using the above-mentioned method, it is possible to define coordinates of the mutual position of a target and indicator in successive moments of the attacking process.

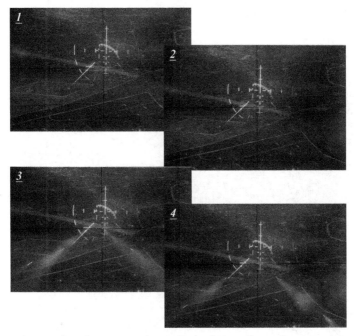

Fig. 6. Photos taken with a photo-control apparatus during the realization of the attacking process with the use of non-guided missiles

Based on the obtained data, it was possible to determine the aiming indicator path relative to a target. Figure 7 depicts the path. When analyzing the position of the central point of the aiming indicator, we can assume that the position adopting the chaotic motion of the indicator was the proper position that completely reflects the nature of the real process.

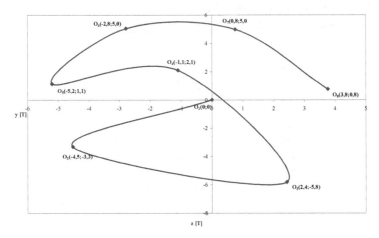

Fig. 7. The course of changes in the position of the aiming indicator relative to a target during the realization of the aiming process with the use of non-guided missiles.

The variance values were determined for the data presented in Fig. 7. The values are as follows:

$$\sigma_z^2 = 14{,}24 \left[T^2 \right], \qquad \sigma_y^2 = 22{,}80 \left[T^2 \right] \tag{72}$$

By substituting the above equation values (58) and on the basis of the recorded data, it was possible to determine a graphical form of the probability density function (Fig. 8) that characterizes the concurrence of the aiming indicator with a target during the execution of the aiming process.

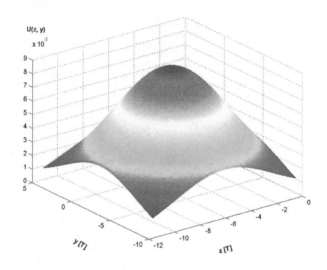

Fig. 8. A graph of the probability density function of indicator deviations during the execution of the aiming process with the use of non-guided missiles

6. Summary

Works carried out during the process of maintaining aim to ensure the required level of safety concerning aircraft engineering and to maintain it in good working condition. This is achieved by carrying out planned works and systematic checks of diagnostic parameter values. Apart from identification, diagnostic testing includes two more aspects concerning the technical state genesis and prediction. That is why, for safety and reliability reasons, it is important to develop methods enabling prediction of the technical state of devices on the basis of information obtained during the maintenance process. The 4rd chapter comprises the presentation of the probabilistic method for the determination of residual durability of devices on the basis of their diagnostic parameter changes registered during the process of maintaining. The application of the above-mentioned method may facilitate the military aircraft maintenance process by limiting the number of stoppages through the indication of a time of next maintenance works for a specified device/system. It shall be emphasized that the presented method is universal as it can be applied to the maintenance process modernization not only in respect of aircraft engineering but also in respect of any field where device/system diagnostic parameters are registered.

The process of operating is inevitably connected with "an operational effect" which results from the completion of a particular combat mission. Depending on a combat mission, this effect will concern, for example hitting the target, intercepting an enemy, identifying the target to attack, etc. The operational effect is always obtained during flight. Due to flying conditions of the military aircraft, we can list a number of destructive factors reducing the value of the obtained operational effect. Analyzing the process of operating, we can state that one of the most significant "cells" in this process is the flying military personnel – a pilot. His task involves the appropriate configuration of the military aircraft systems and the performance of the aiming process that generally comes down to the process of making an aiming indicator coincide with a target. The method presented in the 5th chapter enables the quantitative assessment of the aiming process quality. The results obtained in this way and supported by parameters describing conditions in which a combat task was conducted may constitute the basis for the evaluation of the realization of both a current combat task and the progress in training (considering the series of tasks of a given type in a specified time interval).

7. References

Fisz M. (1958). Probability Calculus and Mathematical Statistics. PWN, Warsaw, Poland

Kaczmarski W. (1990). Aircraft Weapons. Part II. Aircraft Sights Handbook. DWL, Poznan, Poland

Moir I.; Seabridge A. (2006). Military Avionics System. Chichester, England: Wiley

Olearczuk E.; Sikorski M.; Tomaszek H. (1978). Aircrafts maintenance. MON, Warsaw, Poland

Skomra A., Tomaszek H.; Wroblewski M. (1999). Tactical and Technical Characteristics and the Effectiveness of Combat Air Munitions. Military Academy of Technology-Textbook, Warsaw, Poland

Su-22M4 Handbook 7. Weapons. Part VII. Technology of Periodic Service Works. DWLiOP, 1986, Poznan, Poland

Tomaszek H.; Wazny M. (2008). The outline of the assessment of durability against surface wear of a construction element with the use of the distribution of time of the exceedence of limit state (admissible state). ZEM, Vol. 3(155) 2008. pp. 47-59, ISSN: 0137-5474, Radom, Poland

Tomaszek H.; Zurek J.; Loroch L. (2004). The outline of a method of estimation reliability and durability of aircraft's structure elements on the basis of destruction process description. ZEM, Vol. 3(139) 2004. pp. 73-85, ISSN 0137-5474, Radom, Poland

Wazny M. (2003). The analysis of operating causes of the dispersion of selected munition and their influence on the air weapons effectiveness. Military Academy of Technology 2003, Warsaw, Poland

Wazny M. (2008). The method of determining the time concerning the operation of a chosen navigation and aiming device in the operation system. Maintenance and Reliability Nr2/2008, 2(38), pp. 4-11. ISSN: 1507-2711, Lublin, Poland

Wazny M.; Wojtowicz K. (2008). The analysis of the military aircraft maintains system and the modernization proposal.: Maintenance and Reliability Nr3/2008, 3(39), pp. 4-11, ISSN: 1507-2711, Lublin, Poland

www.airliners.net

6

Automatic Inspection of Aircraft Components Using Thermographic and Ultrasonic Techniques

Marco Leo
Consiglio Nazionale delle Ricerche- Istituto di Studi sui Sistemi Intelligenti per l'Automazione
Italy

1. Introduction

Safety in aeronautics could be improved if continuous checks were guaranteed during the in-service inspection of aircraft. However, until now, the maintenance costs of doing so have proved prohibitive. In particular, the analysis of the internal defects (not detectable by a visual inspection) of the aircraft's composite materials is a challenging task: invasive techniques are counterproductive and, for this reason, there is a great interest in the development of non-destructive inspection techniques that can be applied during normal routine tests.

Non Destructive Testing & Evaluation (NDT & E) techniques consist of a data acquisition phase (based on any scanning method that does not permanently alter the article being inspected) followed by a data analysis phase carried out by qualified personnel. In particular, transient thermography and ultrasound analysis are two of the most promising techniques for the analysis of aircraft composite materials (Hellier, 2001).

Non-destructive evaluation requires an excessive amount of money and time and its reliability depends on a multitude of different factors. These range from physical aspects of the technology used (e.g., wavelength of ultrasound) to application issues (e.g. probe coupling or scanning coverage) and human factors (e.g. inspector training and stress or time pressure during inspection) (Kemppainen. & Virkkunen, 2011).

Most of the work in the literature concentrates on the study of data acquisition and manipulation processes in order to prove the relationship between data and structural defects or composition of the material (Chatterjee et al., 2011). Unfortunately only some of the work from the literature concentrates on the posterior analysis of the acquired data in order to (fully or partially) delegate, to some computational algorithm, the automatic recognition of material composition, operative conditions, presence of defects, and so on. This is undoubtedly a very attractive research field since it can reduce operational costs, save time and make the process independent from human factors. However, the development of proper algorithms and methodologies is in its infancy and their level of inspection reliability is still inadequate for those sectors (namely, transportation) where an error can have serious health and safety consequences.

The pioneering work on the a posteriori analysis of data dates back to the early 1990s: it suggested that solutions to the problem of automatic ultrasonic NDT data interpretation could be found by expert systems which embody the knowledge of human interpreters (McNab & Dunlop, 1995) (Hopgood et al., 1993) (Avdelidis et al., 2003) (Meola et al., 2006) (Silva et al., 2003). More effective approaches, based on advanced signal processing and artificial intelligence paradigms, have been proposed in the last decade (Benitez et al., 2009) (Wang et al., 2008).

In this chapter, we address the problem of developing an automatic system for the analysis of sequences of thermographic images and ultrasonic signals to help safety inspectors in the diagnosis of problems in aircraft components in all those cases where the defects or the internal damage are not detectable with a visual inspection. In particular thermographic analysis is proposed to automatically discover water insertions whereas ultrasonic inspection aims at revealing solid insertions of brass foil.

The proposed approach considers two main steps for interpreting thermographic and ultrasonic data: in the first step a pre-processing technique is introduced to clean data from noise and to emphasise embedded patterns and the classification techniques used to compare ultrasonic signals and to detect classes of similar points. In the second step two neural networks are trained to extract the information that characterises a range of internal defects starting from ultrasonic and thermographic signals extracted in correspondence to the defective areas. After that the same neural networks are applied to automatically inspect real aircraft components.

Section 2 gives an overview of the proposed approach whereas section 3 and 4 concentrate on the data pre-processing and classification respectively. Finally, section 5 presents the experimental results on real aircraft material and conclusions are derived in section 6.

2. Overview of the system

The proposed system for automatic inspection of aircraft components is schematized in figure 1. The system takes the data extracted by non destructive processes reported in the literature as transient thermography and ultrasound scanning as input.

Transient thermography is a non-contact technique, which uses the thermal gradient variation to inspect the internal properties of the investigated area. The materials are heated by an external source (lamps) and the resulting thermal transient is recorded using an infrared camera. Of course, this kind of analysis is only applicable to materials that have a good thermal conductivity such as metals and carbon composites. Different types of thermal excitation can be used according to the materials and the defects under investigation: for instance uniform heating, spot heating, and line heating.

Ultrasonic inspection uses instead sound signals at frequencies beyond human hearing (more than 20 kHz) to estimate some properties of the irradiated material by analyzing either the reflected (reflection working modality) or transmitted (transmission working modality) signals. A typical ultrasonic inspection system consists of several functional units: pulser, receiver, transducer, and display devices. A pulser is an electronic device that can produce a high-voltage electrical pulse. Driven by the pulser, the transducer generates a high-frequency ultrasonic wave which propagates through the material. In the transmission

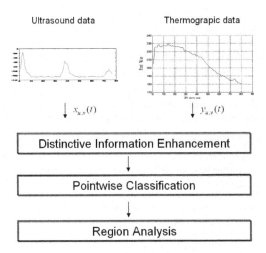

Fig. 1. Scheme of the proposed framework.

modality, the receiver is placed on the opposite side of the material from the pulser, whereas, in the reflection modality, the pulser and the receiver are placed on the same side of the material.

Ultrasonic data can be collected and displayed in a number of different formats. The three most common formats are known in the NDT community as A-scan, B-scan, and C-scan presentations. Each presentation mode provides a different way of looking at and evaluating the region of material being inspected. On the one hand, thermographic analysis is carried out to automatically discover water insertions whereas ultrasonic inspection aims at revealing solid insertions of brass foil.

For thermographic inspection we analyze mono-dimensional signals obtained by considering the time variation of each pixel in the sequence of thermographic images. For each point (i,j) of the material the mono-dimensional signal is generated from the gray levels of the same point in the sequence of images: this signal represents the temperature variation of the material during and after the heating process. This way it is possible to generate spatial-time variant images, the analysis of which allows for the evaluation of the thermal gradient during the heating process.

In Figure 2, the one-dimensional signals extracted from the thermographic sequence of aircraft fuselage are shown: one point belongs to an area affected by the presence of water (red line) whereas the other signal corresponds to non-defective areas (gray lines). From the graph it is clearly evident that a functional description of the intensity variations cannot be easily generalized and the behaviours of points corresponding to defective and non-defective areas are very similar.

For the analysis of ultrasonic data we analyze one-dimensional signals acquired from the reflection working modality and A-scan representation. This means that, for each point of the inspected material, we have a continuous signal that represents the amount of received ultrasonic energy as a function of time.

In figure 3 two ultrasound signals are shown. The signal on top is relative to a non-defective point. Observe that there are large extrema at the beginning and at the end. These changes in ultrasound energy are caused by the transmitted signals being reflected by the boundaries of the material. These boundary extrema are referred to as tool side and bag side peaks, respectively. The ultrasonic signal for an area of material that contains defects is given on the bottom of figure 3. In addition to the boundary extrema, the signals contain extrema at other time locations caused by defective components. The time localization of the additional extrema depends on the defect location in the inspected material.

The temporal evolution of the thermographic and ultrasound signals x(t) is the input to the core of the proposed approach that consists of two main steps: the pre-processing of the data, in order to emphasize the characteristics of the signals belonging to the same class, and the following neural classification.

Pre-processing step allows to discard noise and to enhance the most relevant information for flawed area detection purposes. Two Multi Layer Perceptron (MLP) neural architectures characterized by the presence of an input layer of source nodes, a hidden layer and an output layer, are then used to build an inspection framework that automatically label each signal as belonging to a flawed area or not.

A final connectivity analysis of all the points labelled as belonging to flawed areas is done in order to both discard isolated false positives and to deduce size and shape of the flawed area as a whole.

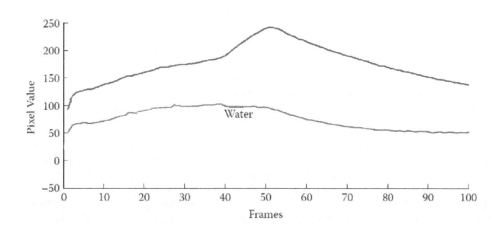

Fig. 2. the one-dimensional signals extracted from the thermographic sequence of aircraft fuselage. The black line corresponds to unflawed areas whereas the red line corresponds to a pixel belonging to water infiltration.

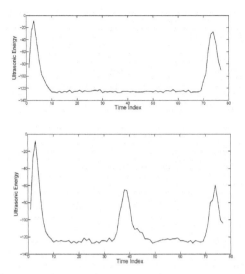

Fig. 3. Two ultrasound signals: the signal on top is relative to a non-defective point. The signal on the bottom is relative to a flawed area.

3. Data pre-processing

The automatic classification of acquired signals as flawed or unflawed is not trivial due to the huge number of intra-class variance: on the one hand ultrasonic and thermal signals relative to unflawed areas can shows different temporal behaviours depending on manufacturing variations in the underlying composite layers or specimen thickness variations. This is evident in figure 4 where different thermographic signals relative to unflawed areas are reported. On the other hand, signals relative to flawed areas can differ since insertions and infiltrations can occur at different locations.

In order to make the classification easier, a pre-processing technique step is then required: on the one hand, it has to increase signal to noise ratio and, on the other hand, to detect and enhance the information that could increase the probability of separating signals belonging to different classes.

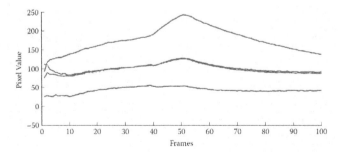

Fig. 4. thermographic signals relative to unflawed areas: their temporal behaviours can strongly differ depending on many factors.

There are many effective signal pre-processing techniques in the literature. Most of them work in a specific domain (time or frequency) whereas a few of them affect both domains simultaneously. In the latter category lies the so-called Wavelet Transform, an extension of Fourier Transform generalized to any wideband transient. For its capability to give a multi-domain representation of the data, the wavelet transform has been used in this work to analyse collected thermographic and ultrasonic data.

In figure 5 the wavelet decomposition (by using Daubechies 3 kernels) at level 3 using a thermographic (top) and ultrasound (bottom) signal are reported.

The next subsection gives some additional theoretical information about the considered pre-processing technique based on Wavelet Transform.

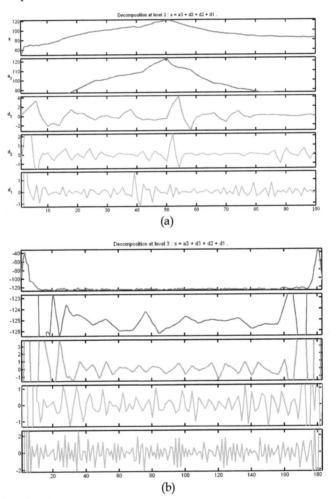

Fig. 5. the wavelet decomposition of a thermographic (on top) and ultrasound (at the bottom) signal.

3.1 Wavelet transform

Let us think about our input as a time-varying signal. To analyze signal structures of very different sizes, it is necessary to use time-frequency atoms with different time supports. The wavelet transform decomposes signals over dilated and translated wavelets Mallat (1999). The signal may be sampled at discrete wavelength values yielding a spectrum. In continuous wavelet transform the input signal is correlated with an analyzing continuous wavelet. The latter is a function of two parameters such as scale and position. The widely used Fourier transform (FT) maps the input data into a new space, the basis functions of which are sins and cosines. Such basis functions are defined in an infinite space and are periodic, this means that FT is best suited to signal with these same features. The Wavelet transform maps the input signal into a new space which basis functions are usually of compact support. The term wavelet comes from well- localized wave-like functions.

In fact, they are well-localized in space and frequency i.e. their rate of variations is restricted.

Fourier transform is only local in frequency not space. Furthermore, Fourier analysis is unique, but wavelet not, since there are many possible sets of wavelets which one can choose.

Our trade-off between different wavelet sets is compactness versus smoothness. Working with fixed windows as in the Short Term Fourier Transform (STFT) may bring about problems. If the signal details are much smaller than the width of the window they can be detected but the transform will not localize them. If the signal details are larger than the window size, then they will not be detected properly. The scale is defined by the width of a modulation function. To solve this problem we must define a transform independent from the scale. This means that the function should not have a fixed scale but should vary. To achieve this, we start from a function $\psi(t)$ as a candidate of a modulation function and we can obtain a family starting from it by varying the scale s as follows:

$$\psi_{s,t}(u) = \psi_s(u\,t) = |s|^p\,\psi\left(\frac{u\,t}{s}\right) = \frac{1}{|s|^p}\,\psi\left(\frac{u\,t}{s}\right)$$

If ψ has width T then the width of ψ_s is sT. In terms of frequencies, the smaller the s the higher the frequencies ψ_s and vice versa.

The continuous wavelet transform $\tilde{X}$ is the result of the scalar product of the original signal $x(t)$ with the shifted and scaled version of a prototype analysing function $\psi(t)$ called mother wavelet which has the characteristic of a band pass filter impulse response.

The coefficients   of the transformed signal represent how closely correlated the mother wavelet is with the section of the signal being analyzed. The higher the coefficient, the more the similarity.

Calculating wavelet coefficients at every possible scale is a fair amount of work, and it generates a great amount of data. If we choose scales and positions based on the power of two (called dyadic scales and positions) then our analysis will be much more efficient. This analysis is called the *discrete wavelet transform*.

In the discrete case, WT is sampled at discrete mesh points and using smoother basis functions. This way a multiresolution representation of the signal $x(t)$ can be achieved.

Notice that the wavelet transform can be written as a convolution product (it is a linear space-invariant filter):

$$\tilde{X}(s,t) = x(u)\psi_{s,t}(u)du = \langle \psi_{s,t}, x \rangle$$

This leads to a fast and efficient implementation of the wavelet transform for a discrete signal obtained using digital filtering techniques. The signal to be analyzed is passed through filters with different cut off frequencies at different scales. The wavelet transform for a discrete signal is computed by successive low-pass and high-pass filtering of the discrete time-domain signal. Many filter kernels can be used for this scope and the best choice depends on the features of the input signal that have to be exploited.

At each decomposition level, the half-band filters produce signals spanning only half the frequency band. This doubles the frequency resolution as the uncertainty in frequency is reduced by half. At the same time, the decimation by 2 doubles the scale. With this approach, the time resolution becomes arbitrarily good at high frequencies, whereas the frequency resolution becomes arbitrarily good at low frequencies.

4. Automatic learning and classification of defective and non-defective patterns

After the pre-processing step the new wavelet based data representations is given as input to an automatic classifier that, after a proper learning phase, is able to label each input stream as belonging to a flawed or unflawed area on the basis of the learned input/output mapping model. One of the most powerful data modelling tools that is able to capture and represent complex input/output relationships is neural network (NN).

4.1 Neural network paradigm

The motivation for the development of neural network technology stemmed from the desire to develop an artificial system that could perform "intelligent" tasks similar to those performed by the human brain. Neural networks resemble the human brain in the following two ways:

1. A neural network acquires knowledge through learning.
2. A neural network's knowledge is stored within inter-neuron connection strengths known as synaptic weights.

The true power and advantage of neural networks lies in their ability to represent both linear and non-linear relationships and in their ability to learn these relationships directly from the data being modelled. Traditional linear models are simply inadequate when it comes to modelling data that contains non-linear characteristics.

The most common neural network model is the multilayer perceptron (MLP), having an architecture as reported in figure 6. This type of neural network is known as a supervised network because it requires a desired output in order to learn. The goal of this type of network is to create a model that correctly maps the input to the output using historical data so that the model can then be used to produce the output when the desired output is unknown.

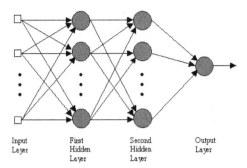

Fig. 6. A feed forward neural network scheme.

The MLP and many other neural networks learn using an algorithm called back propagation. With back propagation, the input data is repeatedly presented to the neural network. With each presentation the output of the neural network is compared to the desired output and an error is computed. This error is then fed back (back propagated) to the neural network and used to adjust the weights such that the error decreases with each iteration and the neural model gets closer and closer to producing the desired output. This process is known as "training".

The hidden layers enable the network to extract higher-order statistics especially when the size of the input layer is large. There is no theoretical limit to the number of hidden layers but, typically, architectures with just one hidden layer are adequate to face the complexity of most of the practical problems . Most used neural architecture have only one hidden layer. Supervised learning involves applying a set of training examples to modify the synaptic weights connecting the neurons of the network. Each example consists of a unique input signal and the corresponding desired response. The network is presented with many examples many times and the synaptic weights are tuned so as to minimize the difference between the desired response and the actual response of the network. The network training is repeated until a steady state is reached, where there are no further significant changes in the synaptic weights.

The input layer has a number of neurons equal to the number of image features. In this work, the features are those extracted after the pre-processing phase. The number of nodes in the output layer depends on the number of classes that the network has to recognize. In our context the network has to recognize the sound point and the defect points (2 output nodes). The number of nodes in the hidden layer is determined by experiment.

There is no quantifiable best answer to the layout of the network for any particular application. There are only general rules picked up over time and followed by most researchers and engineers applying this architecture to their problems.

Rule One: As the complexity in the relationship between the input data and the desired output increases, the number of the processing elements in the hidden layer should also increase.

Rule Two: If the process being modelled is separable into multiple stages, then additional hidden layer(s) may be required. If the process is not separable into stages, then additional layers may simply enable memorization of the training set, and not a true general solution effective with other data.

Rule Three: The amount of training data available sets an upper bound for the number of processing elements in the hidden layer(s). To calculate this upper bound, use the number of cases in the training data set and divide that number by the sum of the number of nodes in the input and output layers in the network. Then divide that result again by a scaling factor between five and ten. Larger scaling factors are used for relatively less noisy data. If you use too many artificial neurons the training set will be memorized. If that happens, generalization of the data will not occur.

5. Experimental setup and results

The composite material used in the experimental tests has an alloy core with a periodic honeycomb internal structure of 128-ply thicknesses (each ply has a thickness of 0.19 mm). The experiments were carried out on two specimens: the first one presents two water infiltrations whereas the second one presents three solid insertions of brass foil (0.02±0.01 mm thickness). One solid insertion was placed two plies from the tool side surface (TOP INSERTION), one at mid part thickness (MIDDLE INSERTION) and the remaining one two plies from the bag side surface (BOTTOM INSERTION). Brass inserts were introduced to represent voids and delamination. In all the cases the defects or the internal damage were not detectable with a visual inspection.

Figure 7 shows the specimens of sandwich material used in the experiments with the graphical information superimposed indicating the exact location of water infiltrations (in blue) and brass foil insertions i.e. top insertion (T) on the left, middle insertion (M) in the centre and bottom insertion (B) on the right.

Fig. 7. the sandwich materials used in the experiments with the superimposed graphical information indicating the exact location of water infiltrations and brass foil insertions

The thermographic image sequence was obtained by using a thermo camera sensitive to the infrared emissions. A quasi-uniform heating was used to guarantee a temperature variation of the composite materials around 20C/sec. In figure 8 one of the thermographic images is reported. Only liquid infiltrations become visible due to the larger thermal variation of the water with respect to solid insertions.

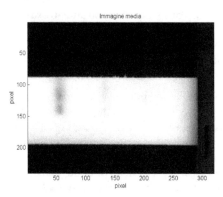

Fig. 8. One of the thermographic images where the liquid infiltrations are visible.

Ultrasonic data were obtained by an ultrasonic reflection technique that uses a single transducer serving as transmitter and receiver (5MHz).

In figure 9, the signal on the left is relative to a non-defective area whereas the signal on the right is relative to a brass insertion placed in the middle of the material thickness and for this reason the corresponding extrema is far from the boundary ones. The signal in the centre of figure 3 is relative to a brass insertion placed very close to the inspected material surface and then the corresponding extrema is mixed with the tool side one. This shows that defective and non-defective areas can have very similar temporal behaviours under ultrasound scanning and this causes traditional NDT techniques to fail.

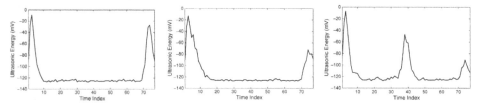

Fig. 9. three ultrasound signals relative to the non-defective area (on the left), a brass insertion placed in the middle of the material thickness (on the right) and very close to the inspected material surface (in the centre).

Acquired experimental data were then represented in the wavelet domain by using Daubechies 3 family of filters and the derived coefficients were given as input to two different neural networks in order to specialize each of them to recognize water infiltration and solid insertion respectively. The defect segmentation step is performed by using neural networks with two output neurons. Each available signal is fed into the net, which classifies it as either relative to defective areas or an unflawed area.

Preliminary experiments aimed at defining the best data model through the selected neural paradigm. In particular they allow the definition of the best number of neurons in the hidden layer and the most suited number of training points. To accomplish this fundamental task different set training examples were built. In particular, for each neural network 3 different training sets consisting of 40, 60 and 80 examples (50% corresponding to unflawed and 50% to flawed areas) were used. At the same time different test sets of points were built for each specimen. In particular, for the specimen with water infiltration two data were built: the first set contained 250 signals relative to unflawed points, the second set contained 250 signals relative to defective areas damaged by the water.

Similarly for the specimen with brass foil insertions 4 data sets were built: the first set contained 250 signals relative to unflawed points, the second set contained 250 signals relative to the defective area corresponding to the brass foil positioned two plies from the tool side surface (Top Insertion), the third set contained 250 signals relative to the defective area corresponding to the brass foil positioned at mid part thickness (Middle Insertion) and finally the fourth set contained 250 signals relative to the defective area corresponding to the brass foil positioned two plies from the bag side surface (Bottom Insertion).

In each experiment a training set was selected and the learned network was then used to classify the data in the corresponding test set. The set of training examples consisted of input-output couples (input signal, corresponding desired response). During the training phase the points of known examples were extracted from the considered materials and continuously fed into the net so that the synaptic weights were tuned to ensure the minimum distance between the actual and the desired output of the net.

Training continues until a steady state is reached, i.e., no further significant change in the synaptic weights could be made to improve net performance. This is repeated also using different configurations of the hidden layer. In particular a number of hidden neurons ranging from 20 to 100 were considered.

The results of this demanding experimental phase are summed up in figure 10 and figure 11.

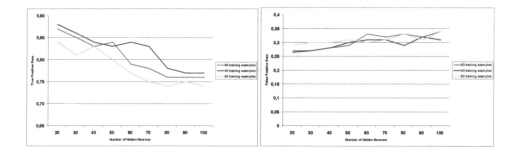

Fig. 10. Experiment results for water infiltration detection using thermal signals when a different number of training examples and hidden neurons were considered

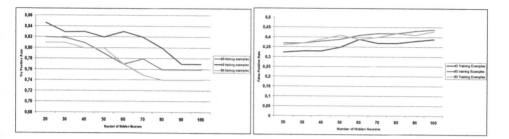

Fig. 11. Experiment results for solid insertion detection using ultrasound signals when a different number of training examples and hidden neurons were considered

Experiments demonstrated that a lower number of hidden layer nodes (i.e 20-30) is a good choice since a larger number of nodes in the inner layer for such situations can drive the classification model to over-fit the training data and to produce a very high failure score. At the same time, experimental proofs pointed out that a limited number of training points (i.e. 40) is the best choice in term of correct classification rate. In other words this is the minimum number of training examples to allow a proper learning of the data distribution and, at the same time, it is the maximum number to avoid data over-fitting case, i.e. to preserve the fundamental capability to classify unknown data (generalization capacity) .

For a better comprehension of experimental results tables I and II report the scatter matrices relative to the experiments performed by using the best network and training set configuration.

	Unflawed	Water insertion
Unflawed	**184/250 (73,6%)**	66/250 (26,4%)
Water insertion	28/250 (11,2%)	**222/250 (88,8%)**

Table I: Scatter matrix derived in the experiments for water infiltration detection.

	Unflawed	Solid Insertion
Unflawed	**169/250 (67,6%)**	81/250 (32,4%)
Brass Foil (top)	36/250 (14,4%)	**214/250 (85,6%)**
Brass Foil (middle)	15/250 (6,0%)	**235/250 (94,0%)**
Brass Foil (bottom)	64/250 (25,6%)	**186/250 (74,4%)**

Table II. Scatter matrix derived in the experiment 1 for brass fail insertion detection.

Table I and II give a quantitative evaluation of the possibility to automatically detect both liquid and solid insertions in composite materials by using thermal and ultrasonic techniques in combination with neural approaches.

In particular, Table II illustrates that brass foil insertions at the mid-thickness level were always better classified than those located either at the top or at the bottom. The defect location is one of the most important factors in ultrasound inspection. The defects placed either at the top or at the bottom of the inspecting structure are in general the most difficult to detect since their echo is mixed with the tool face or the bag side echo. On the contrary, defective areas in the mid-part of the material thickness produce a distinct peak in the signal trend that is straightforward to identify.

In the second part of the experimental phase all the signals extracted by the thermographic and ultrasonic analysis were classified by using the neural networks previously learned.

According to the neural network outputs, a binary image is produced containing black points for defective areas and white points for sound areas.

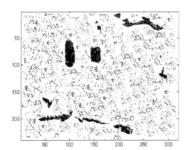 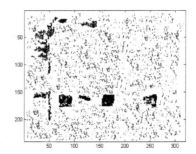

Fig. 12. graphical representation of the raw classification of all the signals extracted from specimens with water infiltration (on the left) and brass foil insertions (on the right).

In figure 12 the graphical representation of the raw classification of all the signals extracted from specimens with water infiltration (on the left) and brass foil insertions (on the right) is reported. Defective areas are correctly detected but there are also a lot of points in the unflawed areas erroneously classified as flawed. For this reason an additional processing step was introduced in order to analyse the output images considering the vicinity of flawed pixels (region analysis). In other words, considering that these false detections were isolated and did not form connected regions having a considerable area value, the elimination of these points was made more straightforward if some a priori knowledge about the minimum expected size of the defective areas is available.

In figure 13 the final outcome is reported after a filtering process based on the connectivity analysis of the detected defective regions and a selection criterion based on removing the regions having an area less than 20 pixels, are shown.

Most of the false flawed points were removed even if some areas in addition to the real defects were still considered flawed. They mainly occurred in correspondence with a variation of the inclination of the surface (see fig. 7): unfortunately, in this unflawed area both thermographic and ultrasound signals changed their slope more evidently with respect to the corresponding signals used to train the net. This problem could be faced by learning the net also on the points belonging to this particular areas. However, this way of proceeding was not considered in this work since it could be counterproductive: the net could miss some real defective areas (or parts of it) and, in our opinion, considering the applicative context, it is critically important to detect all defective points, even at the expense of generating extra false positives

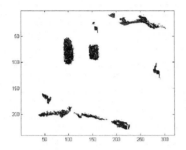 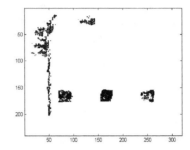

Fig. 13. the result of the cleaning based on the point connectivity analysis on the images reported in figure 8.

6. Conclusion

In this chapter, we address the problem of developing an automatic system for the analysis of sequences of thermographic images and ultrasonic signals to help safety inspectors in the diagnosis of problems in aircraft components.

In particular, thermographic analysis was carried out to automatically discover water insertions whereas ultrasonic inspection aimed at revealing solid insertions of brass foil. Experiments were carried out on real aircraft specimens and demonstrated the capability of the proposed framework to discover flawed areas. A tolerable number of false positive occurrences were also found in correspondence to the part of the specimens having a sloping surface since their points were not included in the learning phase in order to get the best true positive detection rate considering the critical operative context.

Future work will focus on investigating the defect identification capability of the proposed approach. This will be achieved by extending the analysis to material with different thicknesses and different defective insertions. In the future, we will also investigate the possibility of using an unsupervised-learning approach in order to reduce human intervention.

7. References

Kemppainen, M. & Virkkunen, I. (2011). Crack Characteristics and Their Importance to NDE, *Journal of Nondestructive Evaluation*, 2.06.2011 Issn: 0195-9298 Available from http://dx.doi.org/10.1007/s10921-011-0102-z

Chatterjee, K. ; Tuli, S. ; Pickering, S. G. & Almond, D. P. (2011). A comparison of the pulsed, lock-in and frequency modulated thermography nondestructive evaluation techniques, *NDT & E International*, 29.06.2011 Issn 0963-8695 Available from http://www.sciencedirect.com/science/article/pii/S0963869511000892

McNab, A. & Dunlop, I. (1995). A review of artificial intelligence applied to ultrasonic defect evaluation, *Insight*, vol. 37, no. 1, pp. 11–16.

Hopgood, A. A. ; Woodcock, N. ; Hallani, N. J. & Picton, P. (1993). Interpreting ultrasonic images using rules, algorithms and neural networks, *Eur. J. Nondestruct. Test.*, vol. 2, no. 4, pp. 135–149.

Benitez, H. D. ; Loaiza, H. ; Caicedo, E. ; Ibarra-Castanedo, C. ; Bendada, A. & Maldague, X. (2009). Defect characterization in infrared non-destructive testing with learning machines, *NDT & E International*, Volume 42, Issue 7, Pages 630-643, ISSN 0963-8695.

Wang, Y. ; Sun, Y. ; Lv, P. & Wang, H. (2008). Detection of line weld defects based on multiple thresholds and support vector machine, *NDT & E International*, Volume 41, Issue 7, October 2008, pp. 517-524, ISSN 0963-8695.

Hellier, C. (2001). *Handbook of Nondestructive Evaluation*. McGraw-Hill Professional ISBN: 0070281211

Avdelidis, N. P. ; Hawtin, B. C. & Almond, D. P. (2003). Transient thermography in the assessment of defects of aircraft composites, *NDT & E International*, Volume 36, Issue 6, pp. 433-439, ISSN 0963-8695.

Meola, C. ; Carlomagno, G.M., Squillace A. & Vitiello, A. (2006). Non-destructive evaluation of aerospace materials with lock-in thermography, *Engineering Failure Analysis*, Volume 13, Issue 3, pp. 380-388, ISSN 1350-6307

Silva, M. Z. ; Gouyon, R. & Lepoutre, F. (2003) Hidden corrosion detection in aircraft aluminum structures using laser ultrasonics and wavelet transform signal analysis, *Ultrasonics*, Volume 41, Issue 4, pp. 301-305, ISSN 0041-624X.

Part 3

Miscellaneous Topics

7

Review of Technologies to Achieve Sustainable (Green) Aviation

Ramesh K. Agarwal

Department of Mechanical Engineering and Materials Science
Washington University in St. Louis, St. Louis, MO,
USA

1. Introduction

Among all major modes of transportation, people travel by airplanes and automobiles continues to experience the fastest growth. As shown in Figure 1 [1], the travel as measured by Passenger - Kilometers (PKM) is forecasted to more than double from the current 2010 level of ~ 40 trillion PKM to approximately 103 trillion PKM by 2050. Among these two modes of transportation, air travel is experiencing the faster growth. The number of Passenger – Kilometers Travelled (PKT)/ capita by various modes of transportation in different countries is shown in Figures 2(a) - 2(d) [1]. Figures 2(a) and 2(c) also show that the use of personal vehicles compared to public transport (in PKT) is highest in U.S. followed by the wealthier nations. Furthermore, as the per capita income of a nation increases, the travel demand will increase (Figure 3) [1] resulting in greater demand for personal vehicles as well as for air transportation as shown in Figure 1. These projections are based on 3% growth in world Gross Domestic Product (GDP), 5.2% growth in passenger traffic and 6.2% increase in cargo movement. Only major policy changes and intervention by governments through development of infrastructure for public transportation is likely to slow down these trends shown in Figure 1. Most of the energy for transportation is currently provided by the fossil fuels (primarily petroleum). Figure 4 shows the oil consumption for transportation in U.S. and its forecast for the future [2]. Figure 5 shows the relative percentage of fuel consumption by various categories of vehicles in U.S [2]. The consequence of burning fossil fuels is well established in their long term impact on climate and global warming due to Greenhouse Gas (GHG) emissions, primary being the CO_2 and NOx. Table I gives the current level of CO_2 emissions worldwide by ground and air transportation [3] and Figure 6 shows the forecast for the future if the current Business as Usual (BAU) scenario continues [3]. The reduction in GHG emissions due to the burning of fossil fuels is the major goal of "Green Transportation." The "Sustainability" goal is to explore both the technological solutions to increase the efficiency of transportation as well as the alternative carbon neutral fuels (e.g. biofuels among others).

2. Sustainable (green) air transportation

Most of the material presented in this section has been taken from the author's William Littlewood Award Lecture [4]. This section provides an overview of issues related

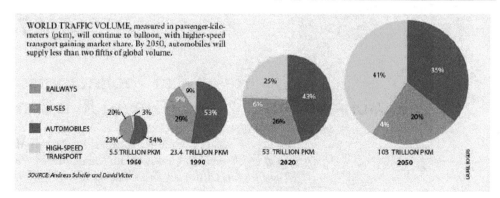

Fig. 1. Global mobility trends from various modes of transportation [1].

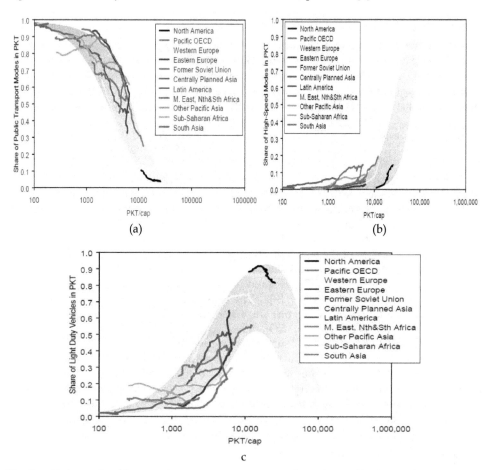

Fig. 2. a: % share of public transport in various countries; b: % share of high speed transport in various countries; c: % share of light-duty vehicle transport in various countries [1].

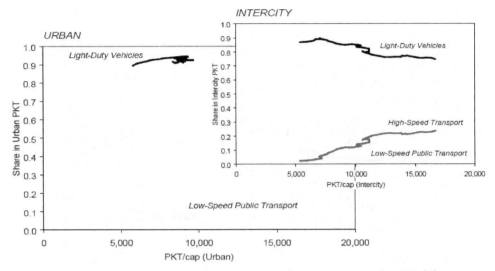

Fig. 2(d). % share of various modes of transportation for inter-city travel in U.S. [1].

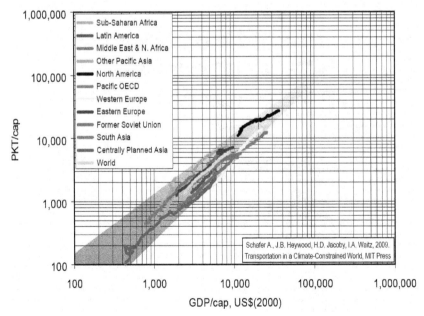

Fig. 3. Travel demand/capita with increase in GDP/capita of nations [1].

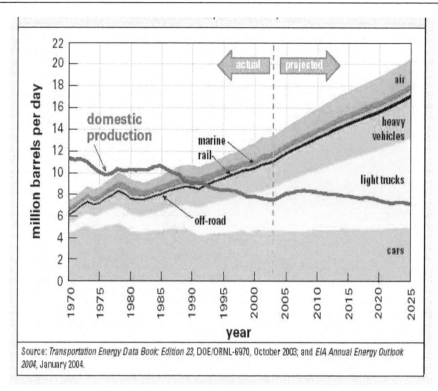

Source: *Transportation Energy Data Book: Edition 23*, DOE/ORNL-6970, October 2003; and *EIA Annual Energy Outlook 2004*, January 2004.

Fig. 4. Fuel consumption in U.S by transport vehicles [2].

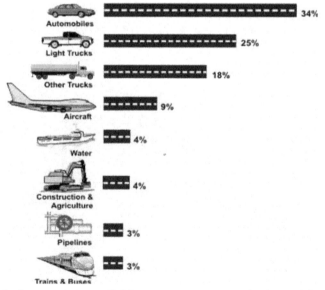

Fig. 5. Relative fuel consumption in U.S by various categories of vehicles [2].

- World Total CO_2 Emissions = 28.4 x 10^9 tonnes (100%)
- US Total CO_2 Emissions = 5.75 x 10^9 tonnes (20.2%)
- China Total CO_2 Emissions = 6.10 x 10^9 tonnes (21.5%)
- World Total from All Transportation = 5.99 x 10^9 tonnes (21.0%)
- World Total from Road Transportation = 3.69 x 10^9 tonnes (13.0%)
- World Total from Air Transportation = 5.68 x 10^8 tonnes (2.0%)
- US Total from Road Transportation ~ 4.46 x 10^9 tonnes (15.6%)
- US Total from Air Transportation ~ 1.39 x 10^9 tonnes (0.5%)

Table 1. Current level of CO_2 emissions from air and ground transportation [3].

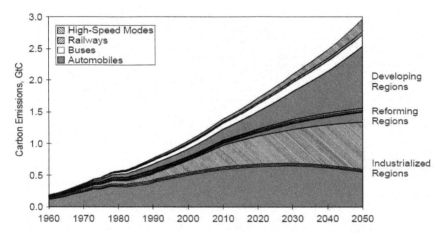

Fig. 6. CO_2 emissions due to world passenger travel in Business as Usual (BAU) scenario [3].

to air transportation and its impact on environment. The environmental issues such as noise, emissions and fuel burn (consumption), for both airplane and airport operations, are discussed in the context of energy and environmental sustainability. They are followed by the topics dealing with noise and emissions mitigation by technological solutions including new aircraft and engine designs/technologies, alternative fuels, and materials as well as examination of aircraft operations logistics including Air-Traffic Management (ATM), Air-to-Air Refueling (AAR), Close Formation Flying (CFF), and tailored arrivals to minimize fuel burn. The ground infrastructure for sustainable aviation, including the concept of 'Sustainable Green Airport Design' is also covered.

As mentioned in the 'Introduction', in the next few decades, air travel is forecast to experience the fastest relative growth among all modes of transportation, especially due to many fold increase in demand in major developing nations of Asia and Africa. Based on these demands for air travel, Boeing has determined the outlook for airplane demand by 2025 as shown in Figure 7 [5]. Figure 8 shows various categories of 27,200 airplanes that would be needed by 2025 [5]. The total value of new airplanes is estimated at $2.6 trillion. As a result of three fold increase in air travel by 2025, it is estimated that the total CO_2 emissions due to commercial aviation may reach between 1.2 billion tonnes to 1.5 billion tonnes annually by 2025 from its current level of 670 million tonnes. The amount of nitrogen oxides around airports, generated by aircraft engines, may rise from 2.5 million tonnes in 2000 to 6.1 million tonnes by 2025. The number of people who may be seriously affected by aircraft noise may rise from 24 million in 2000 to 30.5 million by 2025. Therefore there is urgency to address the problems of emissions and noise abatement through technological innovations in design and operations of the commercial aircraft.

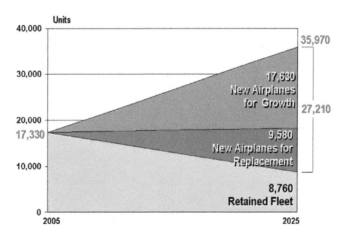

Fig. 7. Boeing market forecast for new airplanes [5].

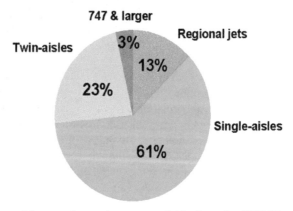

Fig. 8. Boeing demand forecast for various types of Airplanes by 2025 [5].

2.1 Environmental challenges

To meet the environmental challenges of the 21st century, as a result of growth in aviation, the Advisory Committee for Aeronautical Research in Europe (ACARE) has set the following three goals for reducing noise and emissions by 2020; (a) reduce the perceived noise to one half of current average levels, (b) reduce the CO_2 emissions per passenger kilometer (PKM) by 50%, and (c) reduce the NOx emissions by 80% relative to 2000 reference [6]. NASA has similar objectives for 2020 as shown in Figure 9 for N+2 generation aircraft [7]. It is expected that the technology readiness level (TRL) of N+1, N+2 and N+3 generation will be between 4 and 6 in 2015, 2020 and 2030 timeframes respectively. The

NASA definitions of TRL are given in Reference [8]. TRL 4-6 implies that the key technologies readiness will be somewhere between component/subsystem validation in laboratory environment to system/subsystem model or prototyping demonstration in a relevant environment.

CORNERS OF THE TRADE SPACE	N+1 (2015 EIS) Generation Conventional Tube and Wing (relative to B737/CFM56)	N+2 (2020 IOC) Generation Unconventional Hybrid Wing Body (relative to B777/GE90)	N+3 (2030-2035 EIS) Advanced Aircraft Concepts (relative to B737/CFM56)
Noise (cum below Stage 4)	- 32 dB	- 42 dB	better than -71 dB (55 LDN at average boundary)
LTO NOx Emissions (below CAEP 6)	-60%	-75%	better than -75% plus mitigate formation of contrails
Performance: Aircraft Fuel Burn	-33%***	-40%***	better than -70% plus non-fossil fuel sources
Performance: Field Length	-33%	-50%	exploit metro-plex concepts

*** An additional reduction of 10% may be possible through improved operational capability; metro-plex concepts will enable optimal use of runways at multiple airports within the metropolitan area

Fig. 9. NASA subsonic fixed wing system level metric for improving noise, emission and performance using technology & operational improvements [7].

The achievement of these goals will not be easy; it will require the cooperation and involvement of airplane manufactures, airline industry, regulatory agencies such as ICAO and FAA, R & D organizations, as well as political will by many governments and support of public. However, these challenges can be met with concerted efforts as stated beautifully by the Chairman, President and CEO of Boeing Company, W. J. McNerney, "Just as employees mastered "impossible" challenges like supersonic flight, stealth, space exploration and super-efficient composite airplanes, now we must focus our spirit of innovation and our resources on reducing greenhouse- gas emissions in our products and operations."

2.2 A List of new technologies and operational improvements for green aviation

Recently, Aerospace International, published by the Royal Aeronautical Society of U.K., has identified 25 new technologies, initiatives and operational improvements that may make air travel one of the greenest industries by 2050 [9]. These 25 green technologies/concept areas are listed below from Reference [9].

1. *"Biofuels* – These are already showing promise; the third generation biofuels may exploit fast growing algae to provide a drop-in fuel substitute.
2. *Advanced composites* – The future composites will be lighter and stronger than the present composites which the airplane manufacturers are just learning to work with and use.

3. *Fuel cells* - Hydrogen fuel cells will eventually take over from jet turbine Auxiliary Power Units (APU) and allow electrics such as in-flight entertainment (IFE) systems, galleys etc. to run on green power.

4. *Wireless cabins* – The use of Wi-Fi for IFE systems will save weight by cutting wiring - leading to lighter aircraft.

5. *Recycling* - Initiatives are now underway to recycle up to 85% of an aircraft's components, including composites - rather than the current 60%. By 2050 this could be at 95%.

6. *Geared Turbofans (GTF)* - Already under testing, GTF could prove to be even more efficient than predicted, with an advanced GTF providing 20% improvement in fuel efficiency over today's engines.

7. *Blended wing body aircraft* - These flying wing designs would produce aircraft with increased internal volume and superb flying efficiency, with a 20-30% improvement over current aircraft.

8. *Microwave dissipation of contrails* – Using heating condensation behind the aircraft could prevent or reduce contrails formation which leads to cirrus clouds.

9. *Hydrogen-powered aircraft* - By 2050 early versions of hydrogen powered aircraft may be in service - and if the hydrogen is produced by clean power, it could be the ultimate green fuel.

10. *Laminar flow wings* – It has been the goal of aerodynamicists for many decades to design laminar flow wings; new advances in materials or suction technology will allow new aircraft to exploit this highly efficient concept.

11. *Advanced air navigation* - Future ATC/ATM systems based on Galileo or advanced GPS, along with international co-operation on airspace, will allow more aircraft to share the same sky, reducing delays and saving fuel.

12. *Metal composites* - New metal composites could result in lighter and stronger components for key areas.

13. *Close formation flying* - Using GPS systems to fly close together allows airliners to exploit the same technique as migrating bird flocks, using the slip-stream to save energy.

14. *Quiet aircraft* - Research by Cambridge University and MIT has shown that an airliner with imperceptible noise profile is possible - opening up airport development and growth.

15. *Open-rotor engines* - The development of the open-rotor engines could promise 30%+ breakthrough in fuel efficiency compared to current designs. By 2050, coupled with new airplane configurations, this could result in a total saving of 50%.

16. *Electric-powered aircraft* - Electric battery-powered aircraft such as UAVs are already in service. As battery power improves one can expect to see batteries powered light aircraft and small helicopters as well.

17. *Outboard horizontal stabilizers (OHS) configurations* – OHS designs, by placing the horizontal stabilizers on rear-facing booms from the wingtips, increase lift and reduce drag.

18. *Solar-powered aircraft* - After UAV applications and the Solar Impulse round the world attempt, solar-powered aircraft could be practical for light sport, motor gliders, or day-VFR aircraft. Additionally, solar panels built into the upper surfaces of a Blended-Wing-Body (BWB) could provide additional power for systems.

19. *Air-to-air refueling of airliners* - Using short range airliners on long-haul routes, with automated air-to-air refueling could save up to 45% in fuel efficiency.

20. *Morphing aircraft* - Already being researched for UAVs, morphing aircraft that adapt to every phase of flight could promise greater efficiency.
21. *Electric/hybrid ground vehicles* – Use of electric, hybrid or hydrogen powered ground support vehicles at airports will reduce the carbon footprint and improve local air quality.
22. *Multi-modal airports* - Future airports will connect passengers seamlessly and quickly with other destinations, by rail, Maglev or water, encouraging them to leave cars at home.
23. *Sustainable power for airports* - Green airports of 2050 could draw their energy needs from wave, tidal, thermal, wind or solar power sources.
24. *Greener helicopters* - Research into diesel powered helicopters could cut fuel consumption by 40%, while advances in blade design will cut the noise.
25. *The return of the airship* - Taking the slow route in a solar-powered airship could be an ultra 'green' way of travel and carve out a new travel niche in 'aerial cruises', without harming the planet."

Some of the ideas listed above require technological innovation in aircraft design and engines, use of alternative fuels and materials while others require operational improvement. Some concepts such as electric, solar and hydrogen powered aircraft are currently feasible but are unlikely to become viable for mass air transportation by 2050. In what follows, we describe the current levels of noise, CO_2 and NOx emissions due to air transportation and possible strategies for their mitigation to achieve the ACARE and NASA goals.

2.3 Noise & its abatement

Historically, the reduction in airplane noise has been a major focus of airplane manufacturers because of its health effects and impact on the quality of life of communities, especially in the vicinity of major metropolitan airports. As a result, there has been a significant progress in achieving major reduction in noise levels of airplanes in past five decades as shown in Figure 10 [10]. These gains have been achieved by technological innovations by the manufacturers in reducing the noise from airframe, engines and undercarriage as well as by making changes in the operations. Worldwide, there has been ten fold increases in number of airports since the 1970s that now impose the noise related restrictions as shown in Figure 11 [11]. The airports have imposed operating restrictions and also there has been special attention paid to the planning, development and management of airports for sustainability. Since 1980, FAA has invested over $5billion in airport noise reduction.

In recent years, the joint MIT/Cambridge University project on "Silent Aircraft" has produced an innovative aircraft/engine design, shown in Figure 12 that has imperceptible noise outside an urban airport [12]. In order to meet the ACARE and NASA goals of reducing the perceived noise by 50% of the current level by 2020, several new technology ideas are being investigated by the airplane and engine manufacturers to both reduce and shield the noise sources as shown in Figure 13 in the chart by Reynolds [13]. The most promising for the near future are the chevron nozzles, shielded landing gears and the ultra high bypass engines with improved fan (geared fan and contra fan) and fan exhaust duct-liner technology. In addition, new flight path designs in ascent and descent flight can reduce the perceived noise levels in the vicinity of the airports.

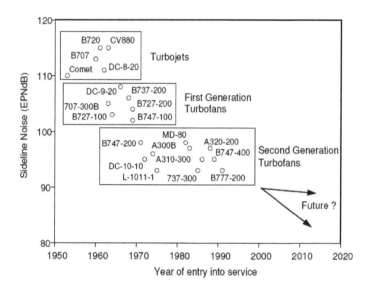

Fig. 10. Reductions in noise levels of aircrafts in past thirty years [11].

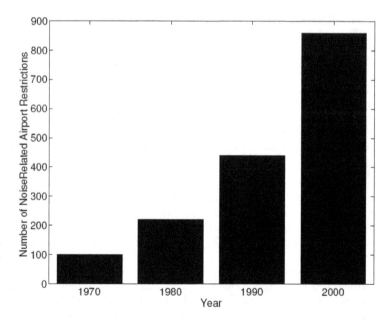

Fig. 11. Number of airports with noise related restrictions in past fifty years [10].

Fig. 12. Silent aircraft SAX – 40: (joint MIT/Cambridge University design) [12].

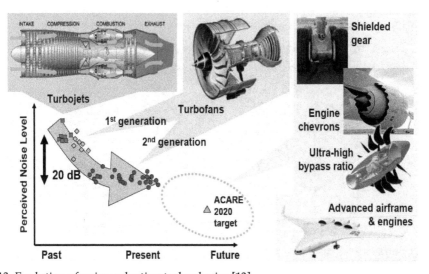

Fig. 13. Evolution of noise reduction technologies [13].

2.4 Emissions and fuel burn

Aviation worldwide consumes today around 238 million tonnes of jet-kerosene per year. Jet-kerosene is only a very small part of the total world consumption of fossil fuel or crude oil. The world consumes 85 million barrels/day in total, aviation only 5 million. At present, aviation contributes only 2-3% to the total CO_2 emissions worldwide [14] as shown in Figure 14. However, it contributes 9% relative to the entire transportation sector. With 2050 forecast of air travel to become 40% of total PKT (Figure 1), it will become a major contributor to GHG emissions if immediate steps towards reducing the fuel burn by innovations in technology and operations, as well as alternatives to Jet-kerosene are not sought and put into effect.

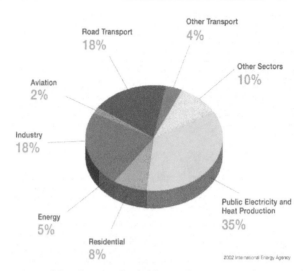

Fig. 14. CO_2 emissions worldwide contributed by various economic sectors [14].

Fig. 15. Contrails & Cirrus Clouds.

Of the exhausts emitted from the engine core, 92% are O_2 and N_2, 7.5% are composed of CO_2 and H_2O with another 0.5% composed of NOx, HC, CO, SOx and other trace chemical species, and carbon based soot particulates. In addition to CO_2 and NOx emissions, formation of contrails and cirrus clouds (Figure 15) contribute significantly to radiative forcing (RF) which impacts the climate change. This last effect is unique to aviation (in contrast to ground vehicles) because the majority of aircraft emissions are injected into the upper troposphere and lower stratosphere (typically 9-13 km in altitude). The impact of burning fossil fuels at 9-13 km altitude is approximately double of that due to burning the same fuels at ground level [15]. The present metric used to quantify the climate impact of aviation is radiative forcing (RF). Radiative forcing is a measure of change in earth's radiative balance associated with atmospheric changes. Positive forcing indicates a net warming tendency relative to pre-industrial times. Figures 16 and 17 show the IPCC (Intergovernmental Panel for Climate Change) estimated increase in total anthropogenic RF due to aviation related emissions (excluding that due to contrails and cirrus clouds) from 1992 to 2050 [16]. It should be noted that in Figures 16 and 17, RF scale is given in W/m^2. It is usually given in mW/m^2; then the numbers in Figures 16 and 17 should be multiplied by 1000 as shown. The horizontal line in Figures 16 and 17 is indicative of the current level of scientific understanding of the impact of each exhaust species.

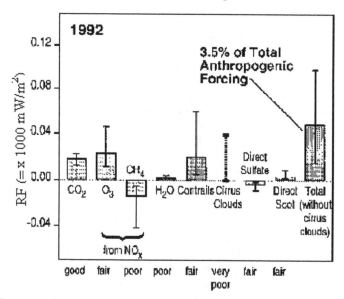

Fig. 16. IPCC estimated Radiative Forcing (RF) due to Emissions – 1992 [16].

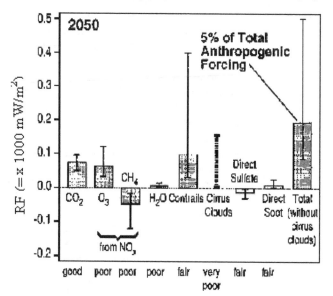

Fig. 17. IPCC estimated Radiative Forcing (RF) due to emissions – 2050 [16].

It should be noted that the RF estimates for 2050 in Figure 17 are based on several assumptions about the growth in aviation, state of technology etc. which are most likely to change. Based on the RF estimates shown in Figures 16 and 17, aviation is expected to account for 0.05K of the 0.9K global mean surface temperature rise expected to occur between 1990 and 2050 [15]. However, RF is not a good metric for weighing the relative importance of short-lived and long-lived emissions. Most importantly, the range of uncertainty about the climate impact of contrails and cirrus cloud remains substantial. According to recent IPCC report, the best estimates for RF in 2005 from linear contrails were 10 (3-30)mW/m^2 and 30(10-80)mW/m^2 from total aviation induced cloudiness, the numbers in bracket give the range of the 2/3 confidence limit [17]. As noted in Reference [17], "the tradeoff estimate of the CO_2 RF in 2000 was 23.5mW/m^2. Despite the growth in CO_2 RF between 2000 and 2005, aviation induced cloudiness remains the greatest contributor to RF according to these estimates. Because of doubts of RF as a metric as well as data spread in cloudiness related RF, the relative contribution of the two (CO_2 and cloudiness) to climate change can not be ascertained with confidence at present time. However, the atmospheric conditions under which an aircraft will generate a persistent contrail – the Schmidt-Appleman criterion [18] – are well understood and can be predicted accurately for a particular aircraft.

Currently there is no technological fix to prevent contrail formation if the atmospheric conditions and engine exhaust characteristics satisfy the Schmidt-Appleman criterion. One assured way of reducing the persistent contrail formation is to reduce aircraft traffic through regions of supersaturated air in which the persistent contrail can form, by flying under, over or around these regions. However, this approach may not be acceptable commercially because of increase in fuel burn, disruption in airline schedule, added ATM workload, and additional operating costs as well as increase in CO_2 and NOx emissions. Because contrail reduction involves an increase in CO_2 and NOx emissions, the best environmental solution is not the complete avoidance of contrails, but a balanced result that minimizes climate impact. This requires a better understanding of the relationship between the properties of the atmosphere (temperature, humidity etc.), the size of the aircraft, the quantity of its emissions (water and particulates), and extent of the persistent contrail and subsequent cirrus formation that results. The adoption of synthetic kerosene produced by Fischer-Tropsch or some similar process offers the prospect of substantial reduction in sulfate and black carbon particulate emissions. This is likely to reduce the extent of contrail and cirrus formation, but the extent of reduction as well as to what extent it would reduce the fuel burn penalty of operational avoidance measures requires further research. Based on the current status, it appears that fuel additives do not offer a significant reduction in contrail formation. The contrail avoidance measures e.g. making modest changes in altitude can reduce contrail formation appreciably with a small penalty in additional fuel burn." Increasing the cruise altitude and higher engine pressure ratio can reduce CO, HC, and CO_2 emissions as well as decrease the fuel burn (improve the fuel efficiency) and facilitate noise reduction. Since higher pressure ratio requires higher flame temperature, the NOx formation rate increases. On the other hand, decreasing the cruise altitude and reducing the engine overall pressure ratio can reduce the NOx but increase the CO_2 emissions. This should be an important consideration in the optimization of future aircraft and engine designs. Research is needed in understanding the impact of cruise altitude on climate. *In addition, there is a need for new optimized aircraft and engine designs that provide a compromise between minimizing the fuel burn and reducing the climate impact.* The lower NOx emissions can possibly be achieved by new combustor concepts such as flameless catalytic combustor and technological improvements in fuel/air mixers using alternative fuels (biofuels), aided by active combustion control. These concepts/technologies should make it possible to meet the N+1 and N+2 generation goals (Figure 9) of achieving the LTO NOx reductions by 60% and 75% respectively below the ICAO standard adapted at CAEP 6 (Committee on Aviation Environmental Protection). It should result in reducing the steepness of the trade-off between NOx and CO_2 emissions and should therefore also help in making a significant contribution to the aircraft performance goal by reducing the fuel burn by 33% and 40% for the N+1 and N+2 generation aircraft respectively. Thus, there are three key drivers in emissions reductions as shown in Figure 18 [19]: (a) innovative engine technologies and aircraft designs, (b) the improvement in ATM and operations, and (c) the alternative fuels e.g. biofuels. The three-prong approach can achieve the goals enunciated by ACARE and NASA by 2020 and beyond. These are discussed in next few sections.

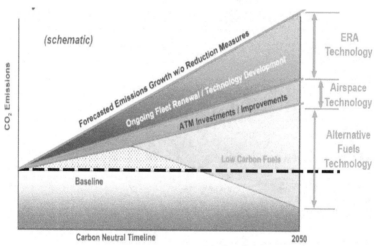

Fig. 18. Key drivers for emissions reductions [19].

2.5 Innovative engine technologies

In cruise condition, the amount of fuel burn varies in inverse proportion to propulsion efficiency and lift-to-drag ratio. Aircraft and engine manufacturers in U.S. and Europe along with several research organizations are developing new engine technologies aimed at improving the propulsion efficiency to reduce the fuel burn and also to simultaneously reduce NOx emissions and noise. The greatest gains in fuel burn reduction in the past sixty years (since the appearance of jet engine) have come from better engines. The earliest engines were turbojets in which all the air sucked in at the front is compressed, mixed with fuel and burned, providing thrust through a jet out the back (see Figure 13). Afterwards, more efficient turbofans were designed when it was realized that greater engine efficiency could be achieved by using some of the power of the jet to drive a fan that pushes some of the intake air through ducts around the core (see Figure 13). Other boosts in efficiency have come from better compressors and materials to let the core burn at higher pressure and temperature. As a result, according to International Airport Transport Association (IATA), new aircraft are 70% more fuel efficient than they were forty years ago. In 1998, passenger aircraft averaged 4.8 liters of fuel/100km/passenger; the newest aircraft – Airbus A380 and Boeing B787 use only three liters. Figure 19 shows the relative improvement in fuel efficiency of various aircraft engines since 1955 [20]. The current focus is on making turbofans even more efficient by leaving the fan in the open. Such a ductless "open rotor" design (essentially a high-tech propeller) would make larger fans possible; however one may need to address the noise problem and how to fit such engines on the airframe. In the short-to-medium-haul market, where most fuel is burned, the open rotor offers an appreciable reduction in fuel burn relative to a turbofan engine of comparable technology, but at the expense of some reduction in cruise Mach number. It is worth noting here that in mid 1980's GE invested significant effort in advanced turbo-prop technology (ATP). The unducted fan (UDF) on a GE36 ultra high bypass (UHB) engine on MD-81 at Farnborough air show in 1988 (Figure 20 [21]) created enormous buzz in the air transportation industry. The author of this paper was at McDonnell Douglas during that period and played a small role in the airframe – engine integration study of MD81 with GE36 ATP. However, in spite of its potential for 30% savings in fuel consumption over existing turbofan engines with comparable performance at speeds up to Mach 0.8 and altitudes up to 30,000 ft, for a variety of technical and business reasons, the advanced turboprop concept never quite got-off the ground [22].

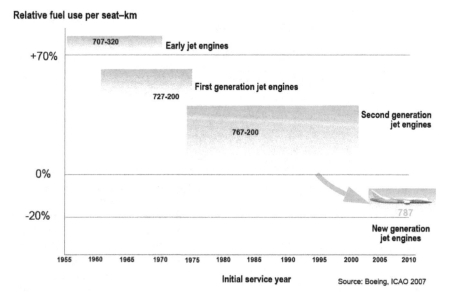

Relative fuel use per seat–km

Fig. 19. Relative improvement in fuel efficiency of various aircraft engines from 1955 to 2010 [20].

Fig. 20. GE36 Turbo-Prop demonstrator engine on MD-81 aircraft [21].

At present in Europe, under the auspices of NACRE (New Aircraft Concept Research Europe), Rolls-Royce and Airbus are making a joint study of the open rotor configurations (Figure 21), including wind-tunnel investigations of power plant installation effects. A key issue in future engine design is how to balance the conflicting aims of reducing fuel burn and NOx emissions (along with the other conflicting aims of reducing noise, weight, initial investment cost and maintenance cost). The results of these types of current and future projects should provide a sounder basis for making decisions between turbofan and open rotor engines for future aircraft. They should also take engine technology well towards its contribution to the goal of a 20% improvement in the installed engine fuel efficiency by 2020.

Fig. 21. Open-Rotor version of pro-active Green Aircraft in NACRE study [17].

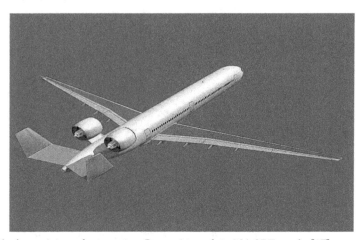

Fig. 22. Turbofan version of pro-active Green Aircraft in NACRE study [17].

2.6 Innovative aircraft designs

As noted in Reference [17], "the classic swept-winged aircraft with a light alloy structure has been evolving for some sixty years and the scope for increasing its lift-to-drag ratio (L/D), if its boundary layers remain fully turbulent, is by now exceedingly limited. Nevertheless, it is well established that increasing L/D is one of the most powerful means of reducing fuel burn. The three ways of increasing L/D are to (a) increase the wing span, (b) reduce the vortex drag factor κ and (c) reduce the profile drag area. The vortex drag factor is a measure of the degree to which the span-wise lift distribution over the wing departs from the theoretical ideal. Current swept-wing aircraft are highly developed and there is little scope for further improvement. A flying wing may enable some additional small reduction in κ, however realistically; there is no real prospect of a significant reduction in fuel burn by altering span-wise loading distributions. Furthermore, increasing the wing span increases wing weight. Current long-range aircraft are optimized to minimize the fuel burn at current cruise Mach numbers. In a successful design the balance between the wing span and wing weight is close to optimum. However, the change to advanced composite materials for the wing structure should result in an optimized wing of greater span; both the B787 and Airbus A350 reflect this. If cruise Mach number is reduced, reducing wing sweep also enables the wing to be optimized at a greater span. The turbofan version of Pro-Active Green Aircraft (Figure 22) included in the NACRE study features a slightly forward swept wing optimized at a significantly higher than usual span. This aircraft is aimed at an appreciable increase in L/D at the expense of some reduction in cruise Mach number. The third option for increasing L/D is to reduce the profile drag of the aircraft. This is seen as the option with the greatest mid-term and long-term potential. For large aircraft, the adoption of a blended wing-body (BWB) layout reduces profile drag by about 30%, providing an increase of around 15% in L/D (estimates of 15% - 20% have been published)." The work on such configurations, both by Boeing (the X-48B, wind tunnel and flight tested at model scale by NASA [Figure 23]) and by Airbus within the NACRE project are proceeding. At present, the first applications of the Boeing BWB are envisaged to be in military roles or as a freighter, with 2030 suggested as the earliest entry to service date for a civil passenger aircraft.

Fig. 23. Boeing/NASA X-48B BWB technology demonstrator aircraft [23].

Fig. 24. Honda Jet [24].

The other well known approach of reducing the profile drag is by the use of laminar flow control in one of its three forms - natural, hybrid or full. Natural laminar flow control was applied with great success in World War II on the P-51 Mustang fighter to give it an exceptional range. As a result there was significant effort devoted to the development of laminar flow airfoils after the end of World War II. In these airfoils, the reduction in friction drag was achieved by moving the transition farther back on the airfoil. In addition, the location of the maximum airfoil thickness was at about 60% of the chord which moved the shock system farther back and reduced the effects of boundary layer thickening and separation caused by it. However in spite of a large number of studies, the success in the laboratory in reducing the drag was never realized on medium size aircraft with swept wings. Therefore, its application has been restricted by a combination of size and wing sweep either to small aircraft with swept wings or medium-sized aircraft with zero or very little sweep. The Pro-Active Green Aircraft in the NACRE project (Figures 21 & 22) is designed to exploit natural laminar flow control and has slightly swept forward wings, to avoid contamination of the flow over the wing by the turbulent boundary layer on the fuselage. "Hybrid laminar flow control employs suction over the forward upper surface of the wing to stabilize the boundary layer. This enables the drag reducing principles that underlie natural laminar flow control to be applied to larger, swept-winged aircraft up to typically the size of the A310. The use of suction to maintain laminar flow over the first half of an airfoil surface has been successfully demonstrated in flight on a B757 wing and an A320 fin. The aerodynamic principles are well understood but the engineering of efficient, reliable, lightweight suction systems requires further work. Thereafter, demonstration of the practicality of the system and assessment of the maintenance and other operational problems that it may encounter will require an extended period of operational validation. The application of suction to maintain laminar flow over the entire surface of a flying wing airliner was proposed by Handley Page in the early 1960s. The proposal was based on the substantial body of research into full laminar flow control, including flight demonstrations, over the preceding decade. Full laminar flow control may have potential to double L/D relative to current standards [17]." Recently unveiled "Honda Jet" (Figure 24) has combined several innovative aircraft and engine design features, namely a combination of over the wing (OTW) engine mount design, natural laminar flow wing (NLF), all composite fuselage, HF – 120 turbofan engine, which give it a 30-35% more fuel efficiency and higher cruise speed than conventional light business jets. This is the range of efficiency that can be achieved for the N+1 generation conventional tube and wing aircraft by 2015. Saeed et al. [25] have recently conducted the conceptual design study of a Laminar Flying Wing (LFW) aircraft capable of carrying 120 passengers. They have estimated that, subject to the constraint of a low cruise Mach number of 0.58, LFC has the potential to reduce aircraft fuel-burn by just over 70%, to about 6 gram per passenger-km (PKM), with a trans-Atlantic range of 4125 nautical miles. Studies of this nature do show the promise of innovative aircraft designs to reduce the fuel burn.

Figure 9 shows the NASA goals of achieving a 33% and 40% reduction in fuel burn for N+1 and N+2 generation aircrafts respectively by using the advanced propulsion technologies, advanced materials and structures, and by improvements in aerodynamics and subsystems. Collier [26] from NASA Langley has provided a detailed outline as to how such savings in fuel burn can be achieved. He has estimated that for a N+1 generation conventional small twin aircraft (162 passengers and 2940nm range), 21% reduction in fuel burn can be achieved by using advanced propulsion technologies, advanced materials and structures, and by improvements in aerodynamics and subsystems. For an advanced small twin, additional 12.3% savings in fuel burn can be achieved by using hybrid laminar flow control as shown in Figure 25.

For a N+2 generation aircraft (300 passengers and 7500 nm range) flying at cruise Mach of 0.85, 40% saving in fuel burn relative to baseline B777-200ER/GE90 can be achieved by a combination of hybrid wing-body configuration (with all composite fuselage), advanced engine and airframe technologies, embedded engines with BLI inlets and laminar flow as shown in Figure 22 [24]. For the baseline aircraft, the fuel burn at Mach 0.85 with 300 passengers for a 7500nm mission range is 237,000 lbs. The N+2 generation aircraft should require 141,100lbs of fuel. As discussed in next few sections, additional savings of 10% in fuel burn can be achieved by operational improvements.

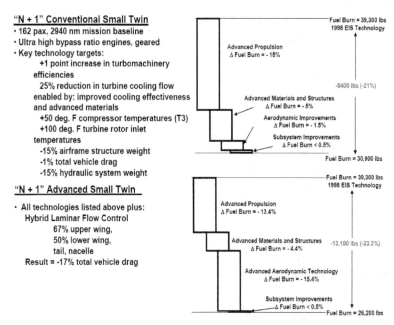

Fig. 25. Reduction in fuel burn for N+1 generation aircraft relative to baseline B737/CFM56 using advanced technologies [26].

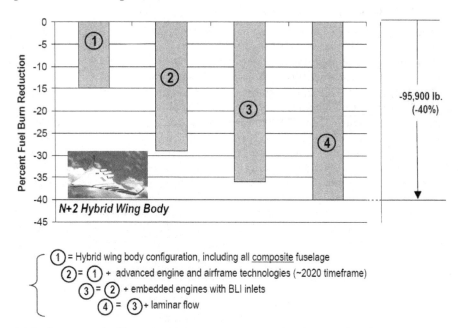

Fig. 26. Reduction in fuel burn for N+2 generation aircraft relative to baseline B777-200ER/GE96 using advanced technologies [26].

2.7 Operational improvements/changes

2.7.1 Improvement in air traffic management (atm) infrastructure

There are many improvements in operations that are being introduced, or will be introduced in the relatively near future that can reduce CO_2 emissions significantly. Foremost among these is the reduction of inefficiencies in ATM, which give rise to routes with dog-legs, stacking at busy airports, queuing for a departure slot with engines running, etc. U.S. Next Generation Air Transportation System (NextGen) architecture and the European air traffic control infrastructure modernization program, SESAR (Single European Sky ATM Research Program), are an ambitious and comprehensive attack on this problem. As described in the U.S. National Academy of Science (NAS) report [27], "NextGen is an example of active networking technology that updates itself with real time-shared information and tailors itself to the individual needs of all U.S. aircraft. NextGen's computerized air transportation network stresses adaptability by enabling aircraft to immediately adjust to ever-changing factors such as weather, traffic congestion, aircraft position via GPS, flight trajectory patterns and security issues. By 2025, all aircraft and airports in U.S. airspace will be connected to the NextGen network and will continually share information in real time to *improve efficiency, safety, and absorb the predicted increase in air transportation.*" Here it is worth noting that operational measures, which can apply to almost the entire world fleet, can have a greater impact, sooner, than the introduction of new aircraft and engine technologies, which can take perhaps 30 years to fully penetrate the world fleet.

2.7.2 Air-to-air refueling (aar) with medium range aircraft for long-haul travel

One particular operational measure that has been advocated is the use of medium-range aircraft, with intermediate stops, for long-haul travel. It has been estimated, using a simple parametric analysis, that undertaking a journey of 15,000km in three hops in an aircraft with design range of 5,000km would use 29% less fuel than doing the trip in a single flight in a 15,000km design. Hahn [28] and Creemers & Slingerland [29] have performed analyses to address this issue using sophisticated aircraft design synthesis methods. Hahn [28], analyzing the assessment for a 15,000km journey in one stage or three, predicted a fuel saving of 29%. Creemers & Slingerland [29], considering a B747-400 (range 13,334km) as the baseline long-range aircraft, designed an aircraft with the same fuselage and passenger capacity (420) but for half the design range (6,672km). This aircraft was predicted to do the long-haul journey in two hops with a 27% fuel saving and at a fuel cost of $70 per barrel, a DOC saving of 9%. Nangia [30] has shown that fuel burn savings of as much as 50% were achievable by using a 5,000km design for a 15,000km journey, since a medium range aircraft can carry a much higher share of their maximum payload as passengers. This difference — which appears essentially to be the difference between medium-range single and long-range twin-aisle aircraft — was not a feature of either the study of Hahn [28] or Creemers & Slingerland [29], which used the same fuselage for both long and medium range designs. This highlights the importance of cabin dimensions and layouts in considering future designs in which, both environmentally and commercially, seat-kilometers per gallon becomes an increasingly important objective. The full system assessment of this proposition, using optimized medium-range aircraft needs further investigation. In order to avoid the intermediate refueling stops, air-to-air refueling (AAR) (Figure 27) has been suggested as a means of enabling medium-range designs to be used on long-haul operations. Nangia has now published a number of papers reporting his work on AAR, which indicate substantial fuel burn savings even after the fuel used by the tanker fleet is taken into account [30, 31].

Fig. 27. Air-to-Air Refueling [30].

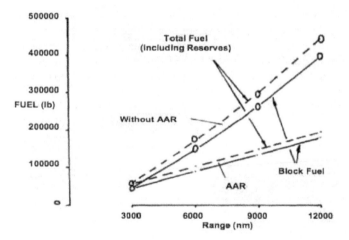

Fig. 28. Savings in fuel burn with Air-to-Air Refuelling (AAR) for long haul flights [31].

Nangia [31] has shown (Figure 28) that an aircraft with L/D = 20, would require 46,147 lbs, 161,269 lbs, and 263,073 lbs of fuel to cover a range of 3,000, 6,000 and 9,000 nautical miles (nm) respectively. With AAR, it will require 92,294 lbs and 138, 441 lbs of fuel for a range of 6,000 and 9,000 nm respectively indicating a savings of 43% and 47% in fuel burn relative to that required without AAR. Accounting for the fuel required by the air tanker – 9,000 lbs for one refueling for a range of 6,000nm and 18,000 lbs for two refueling for a range of 9,000nm, the net savings in fuel burn with AAR are 37% and 41% for a range of 6,000nm and 12,000 nm respectively. However it is paramount that with AAR, the absolute safety of the aircraft is assured.

2.7.3 Close Formation Flying (CFF)

The possibility of using CFF to reduce fuel burn or to extend range is well known. As stated by Nangia [31], "aircraft formations (Figure 29) occur for several reasons e.g. during displays or in AAR but they are not maintained for any significant length of time from the fuel efficiency perspective." The reason is that flying in formation will require extreme safety measures by use of sensors coupled automatically to control systems of individual aircrafts. Furthermore, flying a close formation through clouds or in gusty environment may not be practical. The obvious benefit of flying in formation is a more uniform downwash velocity field, which minimizes the energy transferred into it from propulsive energy consumption. Another benefit is the cancellation of vortices shed from the wing-tips of individual airplanes, except the two outermost ones. How effective this cancellation will be would depend upon the practicality of achievable spacing among the aircrafts. There would also be a substantial benefit in elimination of vortex contrails and cirrus clouds. Recently, NASA conducted tests on two F/A-18 aircraft formations [32]. It was shown that the benefits of CFF occur at certain geometry relationships in the formation, namely the trailing aircraft should overlap the wake of the leading aircraft by 10-15% semi-span in this case. Jenkinson [33] suggested that the CFF of several large aircrafts is more efficient in comparison with flying a very large aircraft. The aircrafts could take-off from different airports and then fly in formation over large distances before peeling off for landing at required destinations. Bower at al. [34] have recently investigated a two aircraft echelon formation and a three aircraft formation of three different aircraft and analyzed the fuel burn. Their study determined the fuel savings and difference in flight times that result from applying CFF to missions of different stage lengths and different spacing between the cities of origin. For a two aircraft formation, the maximum fuel savings were 4% with a tip-to-tip gap between the aircraft equal to 10% of the span and 10% with a tip overlap equal to 10% of the span. For the three aircraft inverted-V formation, the maximum fuel savings were about 7% with tip-to-tip gaps equal to 10% of the span and about 16% with tip overlaps equal to 10% of the span.

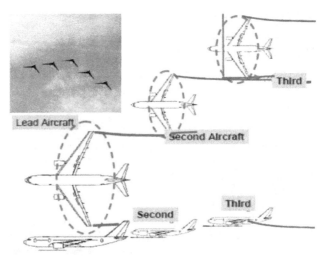

Fig. 29. Three different aircraft type in CFF [31].

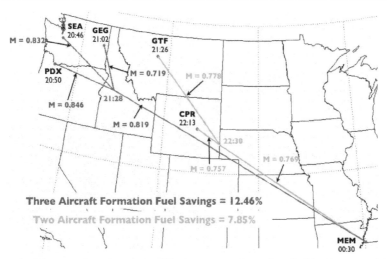

Fig. 30. Five FedEx aircraft in Formation Flight enroute from Pacific Northwest to Memphis [34].

Bower et al. [34] conducted a case study to examine the effect of formation flight on five FedEx flights from the Pacific Northwest to Memphis, TN. The purpose of this study was to quantify the fuel burn reduction achievable in a commercial setting without changing the flight schedule. With tip-to-tip gaps of about 10% of the span it was shown that fuel savings of approximately 4% could be achieved for the set of five flights. With a tip-to-tip overlap of about 10% of the span the overall fuel savings were about 11.5% if the schedule was unchanged. This translated into saving of approximately 700,000 gallons of fuel per year for this set of five flights. Figure 30 shows the three types of aircrafts employed in the study – two Boeing B 727-200, two DC 10-30 and one Airbus A300 – 600F. It should be noted that in CFF, each aircraft will experience off-design forces and moments. It is important that these are adequately modeled and efficiently controlled. Simply using aileron may trim out the induced roll but at the expense of drag. But as Bower et al. [34] have shown, it is possible to realize savings in fuel burn by using the existing aircraft by suitably tailoring the formation.

2.7.4 Tailored arrivals

Boeing [35] is working with several airports, airlines and other partners around the world in developing tools for "tailored arrivals" which can reduce fuel burn, lower the controller workload and allow for better scheduling and passenger connections (Figure 31). To optimize tailored arrivals, additional controller automation tools are needed. Boeing completed the trial of Speed and Route Advisor (SARA) with Dutch air traffic control agency (LVNL) and Eurocontrol in April/May 2009. SARA delivered traffic within 30 seconds of planned time on 80% of approaches at Schiphol airport in Netherlands compared to within 2 minutes on a baseline of 67%. At San Francisco airport, more than 1700 complete and partial tailored arrivals have been completed between December 2007 and June 2009 using the B777 and B747 aircraft. It has been found that tailored arrivals save an average of 950 kg of fuel and approximately $950 per approach. Complete tailored arrivals saved approximately 40% of the fuel used in arrivals. For one year period, four participating airlines saved more than 524,000 kg of fuel and reduced the carbon emissions by 1.6 million kg.

Fig. 31. Airports and Partners participating in the concept of Tailored Arrivals [35].

2.8 Savings in fuel burn by aircraft weight reduction

It is well known that substantial savings in fuel burn can be achieved by reducing the ratio of the empty weight to payload of an aircraft. It can be accomplished by the development and use of lighter and stronger advanced composites, and by reducing the design range and cruise Mach number.

2.8.1 Aircraft weight reduction by use of advanced composites

Reducing the weight of an aircraft is one of the most powerful means of reducing the fuel burn. Boeing and Airbus, as well as other Business and General Aviation aircraft manufacturers are investing in advanced composites which have the prospects of being lighter and stronger than the present carbon fiber composites (CFC). The replacement of structural aluminum alloy with carbon fiber composite is the most powerful weight reducing option currently available to the aircraft designer working towards a given payload-range requirement. The Boeing B787 and Airbus A350 have both taken this step, having wings and fuselage made with CFC. Most new designs are likely to take this path.

2.8.2 Aircraft weight reduction by reducing the design range

Although the historic trend has been in the opposite direction, another powerful means of reducing the weight of an aircraft is to reduce its design range. The study by Hahn [28] has shown that by reducing the design range from 15,000km to 5,000km, with the fuselage and passenger accommodation fixed, it is possible to reduce the operational empty weight (OEW) by 29%. The study by Creemers & Slingerland [29] noted a 17% reduction in OEW by halving the design range from 13,334km to 6,672km. Nangia [30, 31] has also shown that, with the fuselage and number of passengers fixed, wing area increases rapidly to contain the fuel needed and to maintain C_L as the design range increases. Figure 32 shows the aircraft designs and maximum take-off weight MTOW for design range from 3,000 to 12,000 nm. From Nangia's study [31], it is clear that 3,000nm aircraft can provide substantial savings in fuel burn by having less weight and can be used for long range flight by using AAR. In past twenty years, each new aircraft type has achieved 10-15% gain in fuel efficiency. Additional achievements in fuel efficiency by improvements in airframe and engine design will take some time, however, several studies have shown that it is possible to reduce fuel burn significantly by instituting operational measures such as more efficient Air-Traffic Management (ATM), Air-to-Air Refueling (AAR), Close Formation Flying (CFF), Tailored Arrivals, and by reducing the ratio of empty weight to payload.

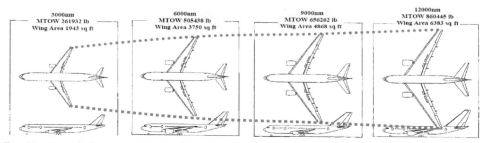

Fig. 32. Aircraft designs, with fixed fuselage, 250 passengers and C_L, for different ranges of operation [30, 31].

2.9 Alternative fuels

All forms of powered ground and air transportation are experiencing the pressure of the need to mitigate greenhouse gas (GHG) emissions to arrest their impact on climate change. In addition the high price of fuel (oil reaching $149/barrel during summer of 2008) as well as the need for energy security are driving an urgent search for alternative fuels, in particular the biofuels. There is emphasis on both the improvements in energy efficiency and new alternative fuels. Aviation is particularly sensitive to these pressures since, for many years, no near term alternative to kerosene has been identified. Until recently, biofuels have not been considered cost competitive to kerosene. An important much desired characteristic of an alternative fuel is whether it can be used without any change to the aircraft or engines. The attractions of such a *drop-in fuel* are clear: it does not require the delivery of new aircraft but the environmental impact of all aircraft flying today can be significantly reduced. Non-drop-in fuels, such as hydrogen or methane hydrates, are unlikely to be used before 2050. The key criteria in identifying that a new alternative fuel would be beneficial in reducing CO_2 emissions should be based on the life cycle analysis of CO_2; the life-cycle CO_2 generation must be less than that of kerosene. Many first generation biofuels have performed poorly against this criterion, though second generation biofuels appear to be far more promising. Furthermore, it is important that there are no adverse side-effects arising from production of the feedstock for biofuel generation, such as adverse impact on farming land, fresh-water supply, virgin rain-forests and peat-lands, food prices, etc. Algae and halophytes (salt-tolerant plants irrigated with sea/saline water) are emerging as potential sustainable feedstock solutions. The alternative fuels need to meet specific aviation requirements and essentially should have the key chemical characteristics of kerosene, that is they won't freeze at flying altitude and they would have a high enough energy content to power an aircraft's jet engine. In addition, the alternative fuel should have good high-temperature thermal stability characteristics in the engine and good storage stability over time.

Interest in biofuels for civil aircraft has increased dramatically in recent years and the focus of the aviation industry on what is and what is not credible in this arena has sharpened. It is clear that a *'drop-in'* replacement for kerosene i.e. the synthetic kerosene appears to be the only realistic possibility in the foreseeable future. The potential of such bio-derived synthetic paraffinic kerosene (Bio-SPK) to reduce the net CO_2 emissions from aviation may well match or exceed that of advances in airframe and engine technologies, and perhaps may achieve reductions across the world fleet sooner than new technologies. In addition, since synthetic kerosene produces substantially less black carbon and sulphate aerosols than kerosene from oil wells, there is a possibility that its use will reduce contrail and cirrus formation as well.

Boeing, Airbus and the engine manufacturers believe that the present engine technology can operate on biofuels (tests are very promising) and that within 5 to 15 years, the aviation industry can convert to biofuels. On 19 June 2009, Billy Glover of Boeing made a presentation to the press at the Paris air show [35] describing the Boeing's "Sustainable Biofuels Research and Technology Program." Tables I and II show the comparisons of key fuel properties of currently used Jet A/Jet A-1 fuel with those with Bio-SPK fuel derived from three different feed-stocks (Jatropha, Jatropha/Algae, and Jatropha/Algae/Camelina) for neat fuel and blends respectively. All Bio-SPK blends met or exceeded the aviation jet fuel requirements. In this presentation, Boeing declared that they are preparing a comprehensive report on Bio-SPK fuels for submittal to ASTM International and expect an approval in 2010. Boeing is working across the industry on regional biofuel commercialization projects. There have already been a few experimental flights operated by several airlines using the biofuel blends and many more are planned in the near future.

Property		Jet A/Jet A-1	ANZ Jatropha	CAL Jatropha/Algae	JAL Jatropha/Algae/Camelina
Freeze Point °C	Max	-40 Jet A -47 Jet A-1	-57.0	-54.5	-63.5
Thermal Stability JFTOT (2.5 hrs. at control temperature) Temperature °C	Min	260	340	340	300
Viscosity -20°C, mm²/s	Max	8.0	3.663	3.510	3.353
Contaminants Existent gum, mg/100mL	Max	7	<1	<1	<1
Metals ppm.	Max	0.1 per metal	<0.1	<0.1	<0.1
Net Heat of Combustion MJ/kg	Min	42.8	44.3	44.2	44.2

Table I. Key Biofuel (Neat) and Jet/Jet A-1 Fuel properties comparison [35].

Property		Jet A/Jet A-1	ANZ Jatropha	CAL Jatropha/Algae	JAL Jatropha/Algae/Camelina
Freeze Point °C	Max	-40 Jet A -47 Jet A-1	-62.5	-61.0	-55.5
Thermal Stability JFTOT (2.5 hours @control temperature)	Min	260	300	300	300
Viscosity -20°C mm2/s	Max	8.0	3.606	3.817	4.305
Contaminants Existent gum, mg/100mL	Max	7	1.0	<1	<1
Net Heat of Combustion MJ/kg	Min	42.8	43.6	43.7	43.5

ANZ = Air New Zealand, CAL = Continental Airline, JAL = Japan Airline

Table II: Key Biofuel (Blend) and Jet/Jet A-1 fuel properties comparison [35].

On 24 February 2008, Virgin Atlantic operated a B747-400 on a 20% biofuel/80% kerosene blend on a short flight between London-Heathrow and Amsterdam. This was the first time a commercial aircraft had flown on biofuel and it was the result of a joint initiative between Virgin Atlantic, Boeing and GE. On 30 December 2008, Air New Zealand (ANZ) conducted a two hour test flight of a B747-400 from Auckland airport with one-engine powered by 50-50 blend (B50) of biofuel (from Jatropha) and conventional Jet-A1 fuel. B50 fuel was found to be more efficient. ANZ has announced plans to use the B50 for 10% of its needs by 2013. The test flight was carried out in partnership with Boeing, Rolls-Royce and Honeywell's refining technology subsidiary UOP with support from Terasol Energy. On January 7th, Continental Airline (CAL) completed a 90-minute test flight using biofuel derived from algae and Jatropha. B737-800 flew from Houston with one engine operating on a 50-50 blend of biofuel and conventional fuel (B50) and the other using all conventional fuel for the purpose of comparison. The biofuel mix engine used 3,600 lbs of fuel compared to 3,700 lbs used by the conventional engine. On January 30, 2009, Japan Airline (JAL) became the fourth airline to use B50 blend of Jatropha (16%), algae (<1%) and Camelina (84%) on the third engine of a 747-300 in one-hour test flight. It was again reported that biofuel was more fuel efficient than 100% jet-A fuel. It should be noted that in all the above demos, biofuel came from sustainable feedstocks (see Tables I and II), sources that neither compete with staple food crops nor cause deforestation. It is worth mentioning that on 1 February 2008, Airbus A380 flew from Filton, U.K. to Toulouse, France with one of its Rolls-Royce engines powered by an alternative, synthetic gas-to-liquid (GTL) jet fuel. Airbus and Qatar Airways are now partners in a GTL consortium which also includes Shell International Petroleum to investigate the use of GTL neat/blend vis-à-vis conventional jet fuel. From an environmental standpoint, it is encouraging and very hopeful that both major manufacturers – Boeing and Airbus are positioning themselves to be at the forefront of alternative and bio-jet fuels. It is surmised that by 2050, with the use of synthetic kerosene derived from biomass, the world fleet CO_2 emissions per passenger-kilometer (PKM) could be lower at least by a factor of three, NOx emissions lower by a factor of 10 and contrail and contrail-induced cirrus formation lower by a factor of 5 to 15.

2.10 Electric, solar or hydrogen powered green aircraft

For many years, there have been several exploratory studies in academia and industry to build and fly aircraft using sources of energy other than Jet-kerosene or synthetic kerosene (biofuels). There have been several success stories in recent years. In March 2008, Boeing successfully conducted a test flight of a manned aircraft powered by PEM hydrogen fuel cells [36], shown in Figure 33. Since fuel cells convert hydrogen directly into electricity and heat without the products of combustions such as CO_2, they use a clean or green source of energy. Fuel cells propelled aircraft is also often called as "an all electric aircraft."

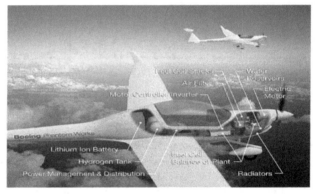

Fig. 33. Boeing PEM Fuel Cell Powered Electric Aircraft [36].

Fig 34. Solar Power Aircraft HB-SIA from SOLAR IMPULSE [37].

Recently in June 2009, the prototype of a new solar-powered manned aircraft was unveiled in Switzerland by the company SOLAR IMPULSE [37]. The airplane is designed to fly both day and night without the need for fuel. The aircraft has a wing span equal to that of a Boeing 747 but weighs only 1.7 tons. It is powered by 12,000 solar cells mounted on the wing to supply renewable solar energy to the four 10HP electric motors. During the day, the solar panels charge the plane's lithium polymer batteries, allowing it to fly at night. To be sure, the fuel-cell propelled electric aircraft and the solar energy driven aircraft are not likely to become feasible for mass air transportation. However, they can become viable for recreation and personal transportation, and possibly as business aircraft in not too distant future. The idea of using liquid hydrogen as a propellant has been around for many decades, but is unlikely to become feasible for commercial aircraft, at least before 2050, because of many challenges that would have to be overcome. Figure 35 shows the artist's rendering of a hydrogen-powered version of A310 Airbus [38]. It is also called a "Cryoplane" because of the very visible cryogenic hydrogen tank located above the passengers. Cryogenic hydrogen is the only possibility for the airplane since the high pressure tanks would be too heavy. The physical properties of the liquid hydrogen determine the appearance of the Cryoplane. Liquid hydrogen occupies 4.2 times the volume of jet fuel for the same energy; therefore the tanks will have to be huge. Jet fuel weighs 2.9 times more than liquid H_2 for the same energy. The reduced weight partly compensates for the increased aerodynamic drag of the tanks. The Cryoplane would have less range and speed than A310. It will have higher empty weight. Furthermore, whatever energy source is used, 30% will be lost in hydrogen liquefaction. In addition, the cost, infrastructure and passenger acceptance issues would have to be addressed. The main advantage of using a hydrogen powered airplane is the reduced emissions as shown in Figure 36 from Penner [39]. Since the use of H_2 does not produce any CO_2, it is dubbed as clean fuel.

Fig. 35. Artist's rendering of a Hydrogen powered version of A310 Airbus [38].

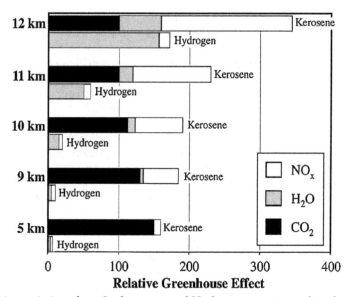

Fig. 36. Relative emissions from Jet-kerosene and Hydrogen at various altitudes [39].

2.11 Modeling environmental & economic impacts of aviation

2.11.1 Cambridge university aviation integrated modeling project (AIM)

Institute for Aviation and the Environment at Cambridge University in U.K. has developed one of the most comprehensive projects – called the Aviation Integrated Modeling (AIM) project to develop a policy assessment capability to enable comprehensive analyses of aviation, environment and economic interactions at local and global levels. It contains a set of inter-linked modules of the key elements which include models of aircraft/engine technologies, air transport demand, airport activity and airspace operations, all coupled to global climate, local environment and economic impact blocks. A major benefit of AIM architecture is the ability to model data flow and feedback between the modules allowing for the policy assessment to be conducted by imposing policy effects on upstream modules and determining the implications through down stream modules to the output metrics, which can then be compared to the baseline case [40].

These modules include: (a) an *Aircraft Technology and Cost Module* to simulate aircraft fuel use, emissions production and ownership/operating costs for various airframe/engine technology evolution scenarios which are likely to have an effect during the period of the forecast; (b) an *Air Transport Demand Module* to predict passenger and freight demand into the future between origin-destination pairs within the global air transportation network; (c) an *Airport Activity Module* to investigate the air traffic growth as a function of passenger and freight growth, to calculate delays and future airline response to them, and to model ground and low altitude operations and congestion to determine LTO emissions as a function of growth in air traffic operations within the vicinity of the airport; (d) an *Aircraft Movement Module* to simulate airborne trajectories between city-pairs, accounting for airspace inefficiencies and delays for given Air Traffic Control (ATC) scenarios and to identify the locations of emissions release from aircraft in flight; (e) a *Global Climate Module* to investigate global environmental impact of aircraft movements in terms of multiple emissions species and contrails; (f) a *Local Air Quality and Noise Module* to investigate local environmental impacts from dispersion of critical air pollutants and noise from landing and take-off (LTO) operations; and (g) a *Regional Economics Module* to investigate positive and negative economic impacts of aviation in various parts of the world, including the increase in direct and indirect employment opportunities in the region. The schematic of the AIM general architecture is shown in Figure 37 [40].

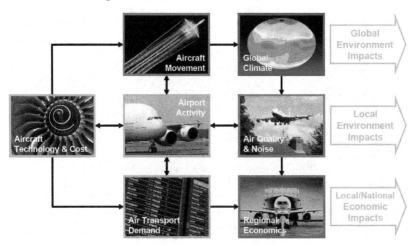

Fig. 37. AIM Architecture [40].

The details of the seven modules and interaction among them are not given here but can be found in many papers listed on the website of the Institute for Aviation and the Environment of Cambridge University in U.K (http://www.iae.damtp.cam.ac.uk/ innovation.html). Here we briefly describe the power of the AIM architecture by reproducing some results from Reynolds et al. [40]. Employing the AIM architecture, Reynolds et al. [40] have performed a case study of the U.S. transportation system, which provides a forecast of air transport passenger demand between 50 major airports in U.S. from 2000 to 2030. The flights between these 50 airports represent over 40% of U.S. scheduled domestic departures in 2000 and nearly 20% the world's scheduled flights. Reynolds et al. [40] conducted simulations under three scenarios: 1. Unconstrained/No Feedback (air transport passenger demands and resulting operations were assumed to grow unconstrained), 2. Feedback of Delay Effects (a simplified airline response to delay is modeled by assuming that the 50% of the cost incurred by the airlines due to delays are passed directly to passengers in the form of higher fares), and 3. Feedback of Delay Effects Plus Per-Km Tax Policy (This is same as scenario 2 , but with a per-Km tax applied to tickets from 2020 onwards with the objective of reducing the Revenue Passenger Km (RPKM) demand in 2020 to 2000 levels, so that the resulting delays and emissions can be directly compared). *Reynolds et al. [40] state that these three scenarios, their associated forecasts and environmental impact results are for illustrative purposes only to show the capabilities of AIM; they do not represent realistic evolutions of the U.S. air transportation system.* The main focus of the scenarios is on interactions between the Air Transport Demand and the Airport Activity Modules. However, one can calculate the en route and local emissions utilizing the capabilities of other modules in AIM integrated structure as given in [40]. Details of the data and assumptions used in the simulation are not presented here. The reader is referred to the paper by Reynolds et al. [40].

Forecasts from 2000 to 2030 for annual demand in terms of Revenue Passenger-Km (RPKM) from the Air Transport Demand Module; and total system aircraft operations, system average arrival delay and local NOx emissions at Chicago O'Hare (ORD) from the Airport Activity Module for the above three scenarios are presented in Figures 38 – 41 from Reynolds et al. [40]. The demand forecasts in Figure 38 include those from Airbus (for U.S market), and Boeing, ICAO and AERO-MS for the North American (NA) market for the purpose of comparison. Since they apply to different route groups and time periods, the start year total RPKM value in each case has been normalized to the historical value for the 50 airports extracted from U.S department of transportation T100 data. Figure 38 shows that for scenario 1, the demand growth measured by increase in RKPM will be 3.5 times the 2000 level by 2030. This is higher than the published estimates as expected given the unconstrained nature of the scenario 1. In scenario 2, the relatively modest feedback of 50% of the increased operating cost to the passenger has a significant effect, particularly over longer time frames. Demand forecast shows a 20% reduction (Figure 38), annual systems operations show a 15% reduction (Figure 39) and average arrival delays show a 50% reduction (Figure 40). Under scenario 3, Figures 38-40 show the effects of distance-based tax; in order to reduce the RPKM demand to 2000 levels in 2020, a 7.7 cents/km charge is required, equating to an additional $300 on a ticket from New York to Los Angeles. Figure 41 shows the annual local emissions at Chicago O'Hare (ORD); all scenarios show an initial gradual increase in emissions which can be explained in conjunction with Figures 38-40 accounting for the increase in RPKM, aircraft operations and arrival delays. The sharp decrease in emissions in scenario 3 in 2020 is due to the reduced operations caused by the introduction of distance-tax policy. The Local Air Quality and Noise Module of AIM architecture can provide results for local air quality at ORD e.g. the annual average NOx concentration at ORD as well as en route CO_2 emissions and global radiative forcing. These results demonstrate that significant insights about environmental and economic impact of aviation can be gained by AIM architecture. It should be noted that many improvements and enhancements to AIM architecture are currently under development at Cambridge.

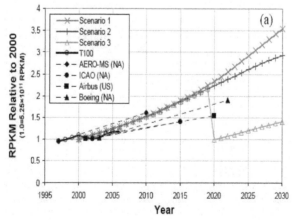

Fig. 38. Forecast of system Revenue Passenger – Km (RPKM) growth at O'Hare [40].

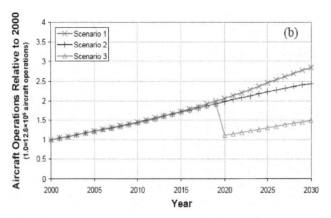

Fig. 39. Forecast of total system aircraft operations at O'Hare [40].

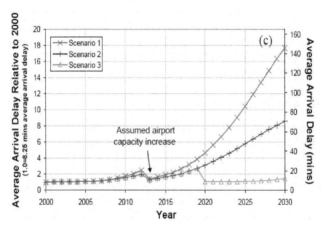

Fig. 40. System average arrival delays at O'Hare [40].

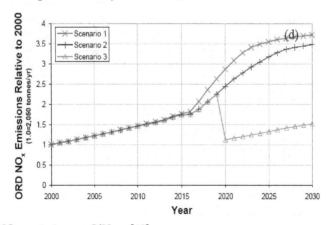

Fig. 41. LTO NOx emissions at O'Hare [40].

2.12 Sustainable airports

The airports and associated ground infrastructure constitute an integral part of Green Aviation. To address the issues of energy and environmental sustainability, the Clean Airport Partnership (CAP) was established in U.S. in 1998 [41] and is the only not-for-profit corporation in the U.S devoted exclusively to improving environmental quality and energy efficiency at airports. CAP believes "that efficient airport operations and sound environmental management must go hand in hand. This approach can reduce costs and uncertainty of environmental compliance; facilitate growth, while setting a visible leadership example for communities and the nation." The airport expansion and the development of new airports should include both the environmental costs and life-cycle costs. Sustainable growth of airports requires that they be developed as inter-modal transport hubs as part of an integrated public transport network. The ground infrastructure development should include low emission service vehicles; LEEDS certified green buildings with low energy requirements, and recyclable water usage. There should be effective land use planning of the area around the airports (including securing land for future development) with active investments into the surrounding communities. Airport expansion must also consider the issue of noise and its impact on the surrounding communities, and should be involved in its mitigation by engaging in the flight path design. The air quality near the airports should be monitored and measures for its continuous improvement should be put in place. In addition, there should be regulatory requirements to set risk limits.

3. Opportunities and future prospects

It is clear that the expected three fold increase in air travel in next twenty years offers enormous challenge to all the stakeholders – airplane manufacturers, airlines, airport ground infrastructure planners and developers, policy makers and consumers to address the urgent issues of energy and environmental sustainability. The emission and noise mitigation goals enunciated by ACARE and NASA can be met by technological innovations in aircraft and engine designs, by use of advanced composites and biofuels, and by improvements in aircraft operations. Some of the changes in operations can be easily and immediately put into effect, such as tailored arrivals and perhaps AAR. Some innovations in aircraft and engine design, use of advanced composites, use of biofuels, and overhauling of the ATM system may take time but are achievable by concerted and coordinated effort of government, industry and academia. They may require significant investment in R&D. It is now recognized by the industry (airlines and manufacturers) as well the relevant government agencies and the policy makers that there is urgent need for action to meet the challenges of climate change; aviation is becoming an important part of it. It is worth noting that in July 2008 in Italy, G8 countries (U.S, Canada, Russia, U.K., France, Italy, Germany and Japan) called for a global emission reduction target of "at least 50%" by 2050, which is in line with goal established by IATA members at their June 2009 Annual General Meeting in Kuala Lumpur, Malaysia. IATA further committed to carbon-neutral traffic growth by 2020. These challenges provide opportunities for breakthrough innovations in all aspects of air transportation.

4. Acknowledgements

The author wants to acknowledge several individuals and sources for their help and permission to use the material from their presentations and papers. The author is grateful to

Dr. Tom Reynolds of the Institute of Aviation and Environment at Cambridge University to allow the use of material used in the section on "Modeling Environmental & Economic Impacts of Aviation." All the material in this section has been taken from Reference [40]. The author is also grateful to Dr. Raj Nangia for helpful discussions and allowing the use of material from several of his papers [30, 31]. Reference [17] has also provided a significant amount of material for several sections. The author is thankful to Dr. Richard Wahls of NASA Langley for his permission to use the material from NASA Langley presentations on 'Environmentally Responsible Aviation.' The author would like to thank Professor Raimo J. Hakkinen for reading the manuscript and for making many helpful suggestions that have improved the paper. Finally, it should be noted that the material used in this review paper has been used from a variety of sources listed in the references; any omission is completely unintentional.

5. References

[1] Schafer, A., Heywood, J.B., Jacoby, H.D., and Waitz, I.A., *Transportation in a Climate Constraint World*, MIT Press, Cambridge, MA, 2009.
[2] Salari, K.,"DOE's Effort to Reduce Truck Aerodynamic Drag Through Joint Experiments and Computations," LLNL-PRES-401649, 28 February 2008.
[3] www.pewclimate.org
[4] Agarwal, R.K., "Sustainable (Green) Aviation: Challenges and Opportunities (2009 William Littlewood Lecture)," SAE Int. J. Aerospace, Vol. 2, pp. 1-20, 2009.
[5] http://www.boeing.com/randy/archives/2006/07/in_the_year_202.html
[6] http://www.acare4europe.com/
[7] NRC Meeting of Experts on NASA's Plans for System-Level Research in Environmental Mitigation, National Harbor, MD, 14 May 2009; Presentation by R.A. Wahls; http://www.aeronautics.nasa.gov/calendar/20090514.htm
[8] Mankins, J.C.,"Technology Readiness Levels," http://www.hq.nasa.gov/office/codeq /trl
[9] Aerospace International, *The Green Issue*, Aerosociety, U.K., March 2009.
[10] Smith, M.J.T., *Aircraft Noise*, Cambridge University Press, Cambridge, U.K., 1989.
[11] Erickson, J.D., "Environmental Acceptability" Office of Environment and Energy, Presented to FAA, 2000.
[12] http://silentaircraft.org/
[13] Reynolds, T.G., "Environmental Challenges for Aviation – An Overview," Presented to Low Cost Air Transport Summit, London, 11-12 June 2008.
[14] www.iea.org
[15] Lee, J.J., Lukachko, S.P., Waitz, I. A., and Schafer, A., "Historical & Future Trends in Aircraft Performance, Cost, and Emissions," Annu. Rev. Energy Environ, Vol. 26, pp. 167-200, 2001.
[16] Penner, J.E., *Aviation and the Global Atmosphere*, Cambridge University Press, Cambridge, U.K., pp. 76-79, 1999.
[17] Royal Aeronautical Society Annual Report, "Air travel - Greener by Design Annual Report 2007-2008," April 2008 (http://www.greenerbydesign.org.uk/).
[18] Schumann, U., "On Conditions for Contrail from Aircraft Exhaust," Meteor. Zeitsch, Vol. 5, pp. 3-22, 1996.
[19] NRC Meeting of Experts on NASA's Plans for System-Level Research in Environmental Mitigation, National Harbor, MD, 14 May 2009; Presentation by A. Strazisar; http://www.aeronautics.nasa.gov/calendar/20090514.htm

[20] www.boeing.com
[21] www.b-domke.de/AviationImages/Propfan/0815
[22] www.flightglobal.com/articles/2007/06/12/214520
[23] http://www.dfrc.nasa.gov/Gallery/Photo/X-48B/HTML/ED08-0092-13.html
[24] http://hondajet.honda.com/
[25] Saeed, T.I, Graham, W.R., Babinsky, H., Eastwood, J.P., Hall, C.A., Jarrett, J.P., Lone, M.M. and Seffen, K.A., "Conceptual Design of a Laminar Flying Wing Aircraft," AIAA 2009-3616, 27th AIAA Applied Aerodynamics Conference, San Antonio, TX, 22-25 June 2009.
[26] Collier, F.S., NASA Langley, "Progress in Environmental Aeronautics," Presentation at Aviation & Environment – A Primer for North American Stakeholders Meeting; http://www.airlines.org/NR/rdonlyres/A78FA93B-986C-4D95-BA87-B4DD961CC369/0/11collier.pdf
[27] National Academy of Science (NAS) Report, "Assessing the Research and Development plan for the Next Generation Air Transportation System: Summary of a Workshop," (http://www.nap.edu/catalog/12447.html), 2008.
[28] Hahn. A.S., "Staging Airliner Service," AIAA 2007-7759, 7th AIAA ATIO Conference, Belfast, 18-20 Sept. 2007.
[29] Creemers, W.L.H. and Slingerland, R., "Impact of Intermediate Stops on Long-Range Jet-Transport Design," AIAA 2007-7849, 7th AIAA ATIO Conference, Belfast, 18-20 Sept. 2007.
[30] Nangia, R.K., "Air to Air Refueling in Civil Aviation," Paper #9, Royal Aeronautical Soc. "Greener by Design" Conference, London, 7 October 2008.
[31] Nangia, R.K., "Way Forward to a Step Jump for Highly Efficient & Greener Civil Aviation – An Opportunity for the Present and a Vision for the Future," Personal Publication RKN-SP-2008-120, September 2008.
[32] Wagner, E., Jacques, D., Blake, W., and Pachter, M., "Flight Test Results for Close Formation Flight for Fuel Savings," AIAA 2002-4490, AIAA Atmospheric Flight Mech. Conf., Monterey, CA, 5-8 August 2002.
[33] Jenkinson, L.R., Caves, R.E, and Rhodes, D.R., "A Preliminary Investigation into the Application of Formation Flying to Civil Operation," AIAA 1995-3898, 1995.
[34] Bower, G.C., Flanzer, T.C. and Kroo, I.M., "Formation Geometries and Route Optimization for Commercial Formation Flight," AIAA 2009-3615, 27th AIAA Applied Aerodynamics Conference, San Antonio, TX, 22-25 June 2009.
[35] Boeing Presentation at Paris Air Show by Billy Glover, June 2009 (http://www.boeing.com/paris2009/media/presentation/june17/glover_enviro_briefing/).
[36] www.boeing.com
[37] www.solarimpulse.com
[38] http://www.planetforlife.com/h2/h2vehicle.html
[39] Penner, J.E., *Aviation and the Global Atmosphere*, Cambridge University Press, Cambridge, U.K., p. 257, 1999.
[40] Reynolds, T.G., Barrett, S., Dray, L.M., Evans, A.D., Kohler, M.O., Morales, M.V., Schafer, A., Wadud, Z., Britter, R., Hallam, H., and Hunsley, R., " Modeling Environmental & Economic Impacts of Aviation: Introducing the Aviation Integrated Modeling Project," AIAA 2007-7751; 7th AIAA Aviation Technology, Integration and Operations Conference, Belfast, 18-20 Sept. 2007.
[41] http://www.cleanairports.com

8

Avionics Design for a Sub-Scale Fault-Tolerant Flight Control Test-Bed

Yu Gu[1], Jason Gross[2], Francis Barchesky[3],
Haiyang Chao[4] and Marcello Napolitano[5]
West Virginia University
USA

1. Introduction

The increasingly widespread use of Unmanned Aerial Vehicles (UAVs) has provided researchers with platforms for several different applications:

1. For carrying remote sensing or other scientific payloads. Highly publicized examples of such applications include the forest fire detection effort jointly conducted by NASA Ames research centre and the US Forest Service (Ambrosia et al., 2004), and the mission into the eye of hurricane Ophelia by an Aerosonde® UAV (Cione et al., 2008);
2. For evaluating different sensing and decision-making strategies as an autonomous vehicle. For examples, an obstacle and terrain avoidance experiment was performed at Brigham Young University to navigate a small UAV in the Goshen canyon (Griffiths et al., 2006); an autonomous formation flight experiment was performed at West Virginia University (WVU) with three turbine-powered UAVs (Gu et al., 2009);
3. As a sub-scale test bed to help solving known or potential issues facing full-scale manned aircraft. For example, a series of flight test experiments were performed at Rockwell Collins (Jourdan et al., 2010) with a sub-scale F-18 aircraft to control and recover the aircraft after wing damages. Another example is the X-48B blended wing body aircraft (Liebeck, 2004) jointly developed by Boeing and NASA to investigate new design concepts for future-generation transport aircraft.

Each of these applications poses different requirements on the design of the on-board avionic package. For example, the remote sensing platforms are often tele-operated by a ground pilot or controlled with a Commercial-off-the-Shelf (COTS) or open-source autopilot (Chao et al., 2009). The UAVs for sensing and decision making research often requires a higher level of customization for the avionic system. This can be achieved through either

---

[1] Research Assistant Professor, Mechanical and Aerospace Engineering (MAE) Department, West Virginia University (WVU), Morgantown, WV 26506, Email: Yu.Gu@mail.wvu.edu;
[2] Ph.D., MAE Dept., WVU, now at Jet Propulsion Laboratory, Pasadena, CA, Email: Jason.Gross@jpl.nasa.gov;
[3] M.S. Student, MAE Dept., WVU, Email: fjbarchesky@gmail.com;
[4] Post-Doctoral Research Fellow, MAE Dept, WVU, Email: Haiyang.Chao@mail.wvu.edu;
[5] Professor, MAE Dept., WVU, Email: Marcello.Napolitano@mail.wvu.edu.

augmenting a COTS autopilot with a dedicated payload computer (Miller et al., 2005), or by having an entirely specialized avionics design (Evans et al., 2001). An alternative approach for smaller UAVs is to instrument an indoor testing environment (How et al., 2008) for measuring aircraft states so that a less complex avionic system could be used on-board the aircraft.

The avionic systems for sub-scale aircraft aimed at improving the safety of full-scale manned aircraft have a different set of design requirements. In addition to providing the standard measurement and control functions, the avionic system also needs to enable the simulation of different aircraft upset or failure conditions. Two general approaches have been used by different research groups. The first approach is to develop a highly realistic experimental environment in simulating a full-scale aircraft operation. For example, the Airborne Subscale Transport Aircraft Research Test bed (AirSTAR) program at NASA Langley research centre uses dynamically scaled airframe equipped with customized avionics for aviation safety research (Jordan et al., 2006) (Murch, 2008). During the research portion of the flight, the aircraft is controlled by a ground research pilot augmented by control algorithms running at a mobile ground station. An alternative approach is to develop a low-cost and expansible aircraft/avionic system for evaluating high-risk flight conditions (Christophersen et al., 2004).

Sub-scale aircraft have played critical complimentary roles to full-scale flight testing programs due to lower risks, costs, and turn-around time. The objective of this chapter is to discuss the specific avionics design requirements for supporting these experiments, and to share the design experience and lessons learned at WVU over the last decade of flight testing research. Specifically, in this chapter, detailed information for a WVU Generation-V (Gen-V) avionic system design is presented, which is based on an innovative approach for integrating both human and autonomous decision-making capabilities. Due to the high risk and uncertain nature of experiments that explore adverse flight conditions, the avionics itself is designed to reduce the risk of a Single Point of Failure (SPOF). This makes it possible to achieve a reliable operation and seamless flight mode switching. The Gen-V avionics design builds upon several earlier generations of WVU avionics that supported a variety of research topics such as aircraft Parameter Identification (PID) (Phillips et al., 2010), formation flight control (Gu et al., 2009), fault-tolerant flight control (Perhinschi et al., 2005), and sensor fusion (Gross et al., 2011).

The rest of the chapter is organized as follows. Section 2 introduces the general design requirements for avionic systems used in fault-tolerant flight control research. Section 3 discusses the overall hardware design architecture and main sub-systems. Section 4 presents the control command signal distribution logic that enables the flexible and reliable transition among different flight modes. Section 5 presents the aircraft on-board software architecture and the real-time Global Positioning System/Inertial Navigation System (GPS/INS) sensor fusion algorithm. Ground and flight testing procedures and results for validating avionics functionalities are discussed in Section 6, and finally, Section 7 concludes the chapter.

2. Avionics design requirements for fault-tolerant flight control research

Fault tolerant flight control research pose special challenges for avionics design due to the complex nature of aviation accidents. The occurrence of aviation accidents can be attributed to many factors, such as weather conditions (e.g. icing or turbulence), pilot errors (e.g. disorientation or mis-judgment), air and ground traffic management errors, and a variety of sub-system failures (e.g. sensor, actuator, or propulsion system failures). Furthermore, the introduction of new technologies in aviation systems poses new threats to the safe operation of an aircraft. For example, modern fly-by-wire flight control systems are known to introduce new failure modes due to their dependence on computers and avionics (Yeh, 1998). Increased automation and flight deck complexity could also potentially degrade situational awareness, and require increased and highly aircraft-specific pilot training. These factors could potentially create new failure scenarios that have not yet been recognized as causes of accidents.

2.1 Research requirements

Due to the complexity of aviation accidents, a multi-functional avionics design is needed to support the fault-tolerant flight control research. The most important requirements for such a design include maintaining accurate and timely measurements of aircraft states, having the ability to emulate various aircraft upset or failure conditions, and providing a flexible interface between humans and automatic control systems. A breakdown of more specific avionics requirements for several aviation safety related research topics is summarized in Table 1.

Research Topic	Specific Avionics Requirements
Aircraft modelling with manually injected manoeuvres	High quality sensor measurements; adequate update rate; monitoring of pilot activities; precise time-alignment of all measured channels.
Aircraft modelling with an On-Board Excitation System (OBES)	Ability to automatically apply pre-specified waveform inputs to control effectors.
Failure emulation	Ability to inject and remove simulated aircraft sub-system failures, such as failures in a particular sensor, actuator, or propulsion unit, or in the control command transmission link.
Fault-tolerant flight control (automatic)	Ability to command and reconfigure individual aircraft control effectors; having low system latency and abundant computational resources.
Fault-tolerant flight control (pilot-in-the-loop)	Ability to augment the pilot command with automatic control algorithms.

Table 1. Design requirements for typical fault-tolerant flight control research topics.

2.2 Operational scenarios

A fundamental difference between operating a sub-scale and a full-scale aircraft is the absence of humans on-board. The removal of the physical presence of human pilots allows the testing of high-risk flight conditions and reduces the cost of the experiment. However, pilots are integral components of modern aviation systems and contributed to 29% of "*fatal accidents involving commercial aircraft, world-wide, from 1950 thru 2009 for which a specific cause is known*" (Planecrashinfo.com, 2011). Pilots are also the ultimate decision-makers on-board; therefore, the evaluations of their response under adverse situations and the detailed understanding of their interaction with the rest of the flight control system play crucial roles in improving aviation safety (NRC, 1997). From this point of view, a realistic fault-tolerant flight testing program should not only take advantage of the low-cost and low-risk features of the sub-scale aircraft, but also to provide a highly relevant operational environment for human pilots. Figure 1 illustrates two potential sub-scale flight testing scenarios for different research topics.

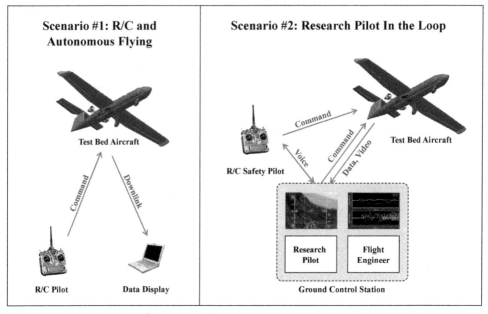

Fig. 1. Two sub-scale aircraft operational scenarios.

Scenario #1 can be used for modelling the aircraft dynamics under different flight conditions and to evaluate automatic Guidance, Navigation, and Control (GNC) algorithms. Within this scenario, a Remote Control (R/C) pilot either directly controls the test bed aircraft or serves as a safety monitor to the on-board flight control system during the test. This scenario provides a simple but reliable method for operating a research aircraft.

Scenario #2 expands upon the first scenario by adding an additional Ground Control Station (GCS), a research pilot, and a flight engineer. The GCS provides a simulated cockpit for the research pilot, who controls the aircraft based on the transmitted flight data and video. This configuration allows the research pilot to have a first-person perspective and enables a fully instrumented flight operation. The role of the flight engineer is to control the configuration of the aircraft by adjusting controller modes/parameters or inject/remove different failure scenarios during the flight. The R/C safety pilot monitors the flight and takes over the aircraft control under emergency situations or during non-research portions of the experiment. Scenario #2 provides additional capabilities for studying the pilot's role in a flight.

2.3 Operational modes

To support the previously described research topics and the two operational scenarios, the following operational modes are typically required:

1. *Manual Mode I – Direct Vision.* An R/C safety pilot has full authority on all control channels in the basic stick-to-surface format. The pilot should always have the option of switching to this mode instantaneously under any conditions as long as the R/C link is available. This mode can be used for aircraft manual take-off and landing, manual PID manoeuvre injection, as well as emergency recovery from other operational modes;
2. *Manual Mode II – Virtual Flight Display.* A research pilot inside the ground control station has full authority on all control channels;
3. *Fully Autonomous Mode.* The on-board flight control system has full control of the aircraft, while the R/C pilot is only serving as an observer and safety backup;
4. *Partially Autonomous Mode.* A subset of the flight control channels is under autonomous control while other channels are still operated by the ground pilot;
5. *Pilot-In-The-Loop Mode.* The pilot command is supplied as input to a Stability Augmentation System (SAS) or a Control Augmentation System (CAS). This mode allows for studying the interaction between a human pilot and the automatic control system;
6. *Failure Emulation Mode.* A simulated failure condition is induced by the on-board computer to one or multiple control channels, while the remaining channels could be under manual, autonomous, or pilot-in-the-loop control;
7. *Fail-Safe Modes.* In the event that the ground pilot could not maintain manual control of the aircraft due to loss of an R/C link, the avionic system should explore redundant communication links and on-board autonomy to help in regaining the aircraft control or minimize the damage of a potential accident.

2.4 Hardware requirement

Due to the high-risk involved in testing various adverse flight conditions and the need for switching between multiple operational modes, the reliability requirements for the avionics hardware are significantly higher than that of a conventional autopilot system for a similar class of UAV. In other words, the avionic system needs to be fault-tolerant itself, and its design should minimize the risk of a SPOF condition. For example, redundant command and control links are needed in case the primary link is lost or interfered. Additionally, the safety pilot should be able to instantaneously switch back to the manual mode from any other operational mode, even in the event of main computer shutdown or power loss.

Additional requirements to the avionics hardware design typically include low-cost, low-weight, low-power consumption, low Electromagnetic Interference (EMI), configurable and expandable, and user-friendly.

3. WVU avionics architecture and main sub-systems

Based on design requirements outlined in the previous section, a Gen-V avionic system is being developed for a WVU 'Phastball' sub-scale research aircraft. The 'Phastball' aircraft has a 2.2 meter wingspan and a 2.2 meter total length. The typical take-off weight is 10.5 Kg with a 3.2 Kg payload capacity. The aircraft is propelled by two brushless electric ducted fans;

each can provide up to 30 N of static thrust. The use of electric propulsion systems simplifies the flight operations and reduces vibrations on the airframe. Additionally, the low time constant associated with an electric ducted fan allows it to be used directly as an actuator or for simulating the dynamics of a slower jet engine. The cruise speed of the 'Phastball' aircraft is approximately 30 m/s. As a dedicated test-bed for fault-tolerant flight control research, the following nine channels can be independently controlled on the 'Phastball' aircraft: left/right elevators, left/right ailerons, left/right engines, rudder, nose gear, and longitudinal thrust vectoring.

The avionic system features a flight computer, a nose sensor connection board, a control signal distribution board, a sensor suite, an R/C sub-system, a communication sub-system, a power sub-system, and a set of real-time software. It performs functions such as data acquisition, signal conditioning & distribution, GPS/INS sensor fusion, GNC, failure emulation, aircraft health monitoring, and failsafe functions. Figure 2 shows the 'Phastball' aircraft along with the main avionics hardware components.

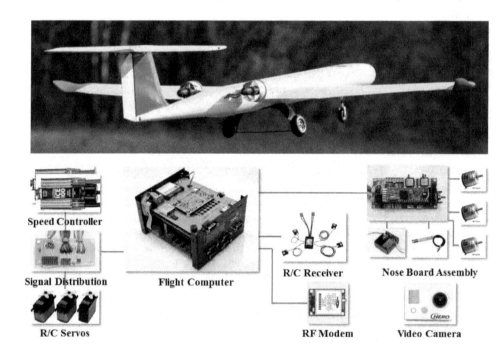

Fig. 2. 'Phastball' aircraft and main avionics hardware components.

A detailed functioning block diagram for the Gen-V avionics hardware design is provided in Figure 3. The functionality of each main sub-system is described in the following sections.

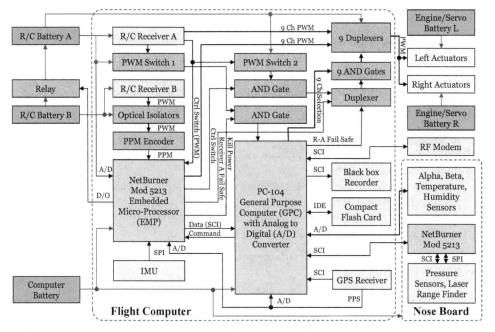

Fig. 3. Functional block diagram for the WVU Gen-V avionics hardware design.

3.1 Flight computer

The Gen-V flight computer integrates the functions of data acquisition, signal conditioning, GPS/INS sensor fusion, failure emulation, automatic control command generation, and control command distribution into a compact package. In terms of hardware, the following main components are included in the flight computer:

1. An Analog Devices® ADIS16405 Inertial Measurement Unit (IMU) that measures the aircraft 3-axis accelerations and 3-axis angular rates. Additionally, it provides readings of the magnetic field for potential use in the navigation filter for an improved aircraft attitude estimation;

2. A Novatel® OEMV-1 GPS receiver that provides aircraft position and velocity measurements. It also provides the precision time information in the form of Pulse Per Second (PPS) signal, which is used to synchronize measurements from different parts of the avionic system;

3. A Netburner® MOD5213 Embedded Micro-Processor (EMP) provides lower level interfaces for measuring human pilot commands in the Pulse-Position Modulation (PPM) format, generating on-board control command in the Pulse-Width Modulation (PWM) format, collecting data from the IMU through a Serial Peripheral Interface (SPI), monitoring battery voltages, and communicating with a general-purpose computer. The EMP also monitors two important PWM signals from the R/C receiver: a *ctrl*-switch and a *kill*-switch. The state of the *ctrl*-switch determines whether the aircraft will be operating in the manual mode or one of the other modes. The *kill*-switch gives pilot the option to power-off the computer during flight if needed for achieving improved

ground-control reliability during the safety critical (such as landing) portion of the flight;

4. Two COTS PWM switches provide independent monitoring of the critical *ctrl*-switch and *kill*-switch;

5. An 800 MHz PC-104+ form factor General-Purpose Computer (GPC) hosts the aircraft on-board software. It also provides additional 16 Analog to Digital Conversion (ADC) channels and 6 Serial Communication Interfaces (SCI) for communicating with the GPS receiver, the EMP, the nose board assembly, and the ground control station;

6. A logic network that distributes control command from both human pilots and automatic control systems to individual actuators based on the selected operational mode of the avionic system;

7. A compact flash memory card storing the operating system, the on-board software, and the collected flight data;

8. A black-box data recorder stores a real-time stream of sensory data, control command, and the avionics health information during the flight.

A detailed description of the aircraft control command generation and distribution is provided in Section 4.

3.2 Sensor suite

In addition to the IMU and the GPS receiver embedded inside the Gen-V flight computer, the '*Phastball*' aircraft is also equipped with three P3America® MP1545A inductive potentiometers for measuring aircraft flow angles, two Sensor Technics® pressure sensors for measuring the dynamic and static pressures, a Measurement Specialities® HTM2500 temperature and relative humidity sensor, and an Opti-Logic® RS400 laser range finder. Additionally, the pilot input, engine operating parameters, and R/C receiver status are also recorded in flight. The aircraft attitude angles are provided with a real-time GPS/INS sensor fusion algorithm, which will be described in Section 5.

3.3 Power system

To reduce SPOF, the arrangement of battery power has been carefully determined. A total of five battery packs are used to power different components of the avionic system. Specifically, an R/C battery-A is connected to R/C receiver-A and an logic network for control command distribution; an R/C battery-B is used to power receiver-B; the computer battery powers EMP, GPC, and all sensing, communication, and data storage devices; engine/servo batteries L and R power the left and right side engines and R/C servos independently. With this configuration, the failure of any given battery would not cause a total loss of aircraft control during the flight. Specifically, if EMP detects that receiver-A battery is low, it activates a relay to tie up R/C batteries A and B so that there would be enough power for a safe landing. If the computer battery loses its power, the logic network powered by receiver-A battery automatically switches to the manual mode and gives the R/C pilot full control authority. If one of the engine/servo batteries fails, the pilot still has independent control for half of the aircraft actuators (propulsion and control surfaces) and would be able to perform a controlled landing.

3.4 Ground control station

The GCS computer collects the aircraft downlink telemetry data, the nose camera video, the weather information, the GPS time/position measurements, the voice communication, as well as inputs from the R/C pilot, the research pilot, and the flight engineer. The data stream is accessible by the research pilot, the flight engineer, and field researchers in near real-time through a local network. For the research pilot station, three displays are provided including an X-Plane® based synthetic-vision primary flight display overlapped with a Heads-Up Display (HUD) that shows the flight parameters and mission constraints, a flight instrumentation display with a navigation window, and a screen showing the real-time flight video transmitted from the aircraft nose camera. The research pilot flies the aircraft through a set of joystick, rudder pedals, and throttle handles. The flight engineer has access to all available flight data and can change the aircraft operational mode or inject/remove failures with or without notifying the research pilot. Figure 4 shows the layout of the GCS vehicle.

 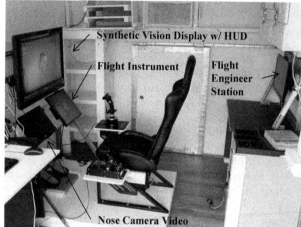

Fig. 4. The exterior (left) and interior (right) of the ground control station vehicle.

The duplex communications between the ground control station and the test bed aircraft is provided with a pair of 900 MHz Freewave ® Radio Frequency (RF) modems. The downlink communication packet contains information about aircraft states and avionics health conditions. The uplink packet integrates both the research pilot control commands and the flight engineer configuration commands. Both the uplink and downlink data are transmitted at a rate of 50Hz.

4. Control command signal distribution

One of the important features of the WVU Gen-V avionics design is its ability to provide a flexible and reliable interface between control commands generated by humans and automatic controllers. This capability is achieved through the interaction of different hardware components and software functions.

4.1 Control command generation

Depending on mission requirements, the aircraft control command could come from several potential sources:

1. *The R/C pilot.* The R/C pilot commands are provided to the flight computer through two redundant R/C receivers (A & B). The antennas of the two receivers are installed at different locations of the aircraft to reduce the likelihood that both are interfered at the same time. Each receiver can operate in either a nominal mode, when maintaining a good reception of the radio signal, or a fail-safe mode, when the communication with the radio transmitter is lost. In the fail-safe mode, each receiver channel output is either independently programmed to a pre-set value, or assigned to latch on to the last received value from the R/C transmitter.

   For R/C receiver-A, 9-channel pilot control commands are sent directly to a duplexer network for later distribution to individual actuators. Two additional channels are used as *ctrl*-switch and *kill*-switch. In order to provide information about both operational modes and the R/C receiver status, the *ctrl*-switch is programmed to have three different output levels: a lower pulse width for '*ctrl*-switch off', a higher pulse width for '*ctrl*-switch on', and a median pulse width indicating the receiver went into a fail-safe mode. A pulse width indicating '*ctrl*-switch on' may trigger a fully autonomous, a partially autonomous, or a pilot-in-the-loop mode depending on additional hardware and software settings.

   The output of receiver-B is first processed with a PPM encoder before being measured with an EMP General-Purpose Timer (GPT). This pilot input is then transmitted to GPC to be used by the flight control system;

2. *The research pilot at GCS.* Commands from the research pilot are transmitted through a pair of RF modems to the flight computer. This signal can be used to control the aircraft directly or indirectly through a SAS or CAS controller;

3. *The Flight Control System (FCS).* The FCS running inside GPC generates the automatic control command based on sensor feedbacks as well as pilot commands provided through either receiver-B (for the R/C safety pilot) or the RF modems (for the research pilot);

4. *Failure Emulation Software (FES).* A faulty actuator locked at a given deflection or a failed engine can both be simulated by sending a constant value to the selected control channel. A slower responding engine can be simulated by inserting additional dynamics between the control command and the engine speed controller. A floating control surface can be simulated with the feedback from a local flow indicator. More complicated failure scenarios can also be introduced through exploring feedbacks from various sensors;

5. *On-Board Excitation System (OBES).* OBES provides specified waveform to be applied on aircraft control actuators. The OBES manoeuvre can be either stand-alone or superimposed onto the pilot or controller commands.

The R/C pilot command is in a PWM format recognizable by R/C servos and engine speed controllers. The commands from the research pilot, FCS, FES, and OBES are first integrated (selected or combined) within the on-board software before being converted into a set of PWM signals. Due to the existence of these two parallel streams of PWM commands, there are several layers of checking and signal distribution to ensure the reliability and the flexibility of the transition.

4.2 Command signal distribution

The command signal distribution system manages and distributes the R/C Pilot Control Command (PCC) provided by receiver-A and the on-board Software-generated Control Commands (SCC) to individual control actuators. Based on the operational mode, the SCC can be one of or a combination of the R/C pilot commands provided by receiver-B, research pilot command, and commands from FCS, FES, and OBES.

The *ctrl*-switch, which the R/C pilot can turn on/off at any given time during the operation, plays a central role in determining the operational mode of the system. Specifically, based on measured receiver-B *ctrl*-switch signal, the EMP sends out a logic (high/low) signal indicating the status (on/off) of the *ctrl*-switch. This status indicator meets with the output of PWM switch-2, which measures the receiver-A *ctrl*-switch signal, at an AND gate. The output of the AND gate, which is called as Confirmed Ctrl Switch Signal (CCSS), becomes logic high only if both input signals are high. This provides a cross-check avoiding accidental activation of the on-board control due to either an EMP or PWM switch-2 failure.

If both receiver-A and B are functioning in the normal mode, a low CCSS initiates the logic network to feed the receiver-A pilot command directly to the control actuators for enabling the pilot manual control. The CCSS can only be overridden in the situation that receiver-A is in the fail-safe mode. Under this condition, the avionic system is able to relay the receiver-B output to actuators through EMP and GPC even if the CCSS signal is low. To achieve this capability, a duplexer is used to switch between CCSS and an EMP provided receiver-A fail-safe indicator. The switching signal for the duplexer is generated by the GPC, which provides a second confirmation that receiver-A is in the fail-safe mode.

To further improve the flexibility of the avionic system and for enabling the partially autonomous mode, another level of logic is provided before the SCC reaches an actuator. Specifically, the GPC is sending out a set of 9-channel selection signals through digital output ports. These channel selection signals are then joined with CCSS at nine AND gates to independently control a 9-channel duplexer network with both SCC and PCC as inputs. Within this configuration, if CCSS is low, all channels will be under manual control. If CCSS is high, the on-board software controls any channel with a high channel selection signal with the rest channels being controlled by the R/C pilot.

The configuration of channel selection signals is normally defined prior to flight based on mission requirements. They can also be modified by the GCS flight engineer during the operation through changing uplink communication packets. Additionally, if receiver-A goes into the fail-safe mode the GPC will activate all channel selection signals along with the fail-safe indicator. This allows the pilot command registered from receiver-B to reach actuators, maintaining the R/C pilot control.

The above-mentioned command signal distribution between PCC and SCC relies on a collaboration of both hardware and software functions. For generating SCC, the integration of commands from R/C pilot, research pilot, FCS, FES, and OBES are performed by the GPC software and are determined based on the specific flight mode. To help clarify the command signal distribution process, pseudo-codes for the EMP software and the command signal distribution portion of the GPC software are provided in Figures 5 and 6 respectively.

```
function EMP_software
      while (1)                                // start an infinite loop
            read IMU data;
            read receiver-B PPM signal;
            if (kill_switch = 'on')
                  set kill_switch pin high;
            else
                  set kill_switch pin low;
            read receiver-A ctrl-switch;
            if (ctrl_switch = 'fail-safe')
                  set fail_safe pin high;
            else
                  set fail_safe pin low;
            if (ctrl_switch = 'on')
                  set ctrl_switch pin high;
            else
                  set ctrl_switch pin low;
            read ADC;
            if (Receiver_A_battery = 'low')
                  tie receiver batteries A&B;
            send data packet to GPC;
            receive control command packet from GPC;
            generate PWM Signal;
```

Fig. 5. Pseudo-code for the embedded micro-processor software.

```
function GPC_Command_Distribution (CC_RCP, CC_GCS, CC_FCS, CC_FES, CC_OBES, CSD,
flight_mode, ctrl_switch, fail_safe)
      // CC - control command, RCP- R/C pilot, GCS- ground control station,
      // FCS - flight control system, FES - failure emulation software,
      // OBES - on-board excitation system, CSD - channel selection data
      if (ctrl_switch = 'off')
            flight_mode= 'Manual I';
      if (fail_safe = 'on')                         // indicating receiver-A fail safe
            set fail-safe pin high;
            flight_mode= 'Fail Safe';
      else
            set fail-safe pin low;
      switch (flight_mode)
            case 'Fail Safe'                  ctrl_command = CC_RCP;
                                              CSD = 511;        // all 9 channels;
            case 'Manual I'                   ctrl_command = CC_RCP;
                                              CSD = 0;    // no channel;
            case 'Manual II'                  ctrl_command = CC_GCS;
            case 'Autonomous'      ctrl_command = CC_FCS;
            case 'Pilot_in_the_loop'          ctrl_command = CC_FCS;
            // the FCS will have pilot input as input in this case.
            case 'Failure Emulation'          ctrl_command = CC_FES;
            case 'OBES'                 ctrl_command = CC_OBES;
            case 'OBES + Manual II'           ctrl_command = CC_GCS+OBES;
            // additional operational modes are available through different
            // combinations of control commands.
      set channel selection digital I/O pins according to CSD;
      send control command packet to EMP;
      return flight_mode;
```

Fig. 6. Pseudo-code for the command signal distribution portion of the GPC software.

5. On-board software

5.1 Operating systems

The use of a general-purpose computer within the avionics design facilitates the use of abundant COTS and open source software products. The on-board Operating System (OS) for GPC is the Linux kernel 2.6.9 patched with Real-Time Application Interface (RTAI) 3.2. An RTAI target was implemented so that Simulink® schemes can be compiled into real-time executable files using the Matlab Real Time Workshop®. The auto-coding capability allows for a rapid integration and testing of algorithms developed by independent researchers.

The NetBurner® MOD5213 EMP uses a µC/OS real-time operating system. The main functionality of the EMP software was outlined in Figure 5.

5.2 GPC software

The GPC software has a modular structure that is first implemented in Simulink® before being compiled into real-time executable files. Each module is either a combination of existing Simulink blocks or a custom S-function written in C language. The modular structure allows for parallel development and debugging, quick and easy configuration for different mission requirements, and intuitive visual interpolation of the software. Additionally, without any modification the same software module can be first simulated in the Simulink® environment before being tested in flight. The main modules of the GPC software and their connectivity are shown in Figure 7.

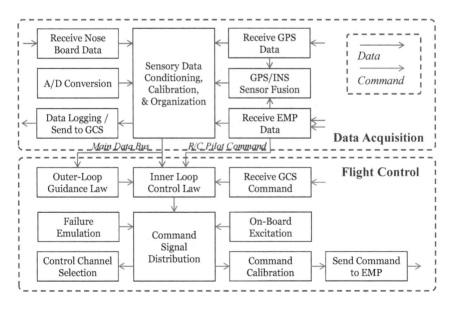

Fig. 7. GPC on-board software architecture.

5.3 GPS/INS sensor fusion

A low-cost INS is regulated with measurements from a GPS receiver to provide navigation solutions to the avionic system. Including a real-time GPS/INS sensor fusion algorithm eliminates the need of heavier and more expensive navigation-grade inertial sensors for a small and low-cost research aircraft.

A 9-state Extended Kalman Filter (EKF) based GPS/INS sensor fusion algorithm is selected for the Gen-V avionics design after a comprehensive comparison study of different sensor fusion formulations and nonlinear filtering algorithms (Rhudy et al., 2011) (Gross et al., 2011). This solution provides a good balance between attitude estimation performance and computational requirements. Within this formulation, the state vector includes the aircraft 3-axis position (x, y, z) and velocity (V_x, V_y, V_z) defined in a Local Cartesian frame (L), and aircraft attitude represented by three Euler angles (φ, θ, ψ) defined in the aircraft Body-axis (B):

$$\mathbf{x} = \begin{bmatrix} x^L & y^L & z^L & V_x^L & V_y^L & V_z^L & \phi^B & \theta^B & \psi^B \end{bmatrix}^T \tag{1}$$

During the state prediction stage, the inertial measurements in terms of three axis accelerations $(\tilde{a}_x^b = a_x^b + v_{ax}, \tilde{a}_y^b = a_y^b + v_{ay}, \tilde{a}_z^b = a_z^b + v_{az})$, and 3-axis angular rates $(\tilde{p}^b = p^b + v_p, \tilde{q}^b = q^b + v_q, \tilde{r}^b = r^b + v_r)$ are integrated to provide an estimate of the state vector $\mathbf{x}$. Each measurement (e.g. $\tilde{a}_x^b$) is a combination of the true measured parameter (e.g. a_x^b) and an noise term (e.g. v_{ax}). The noise is assumed to be zero mean and normally distributed, with its variance approximated by statistical analyses from static ground tests.

The three position states are predicted through straight forward integration, as represented in discrete-time:

$$\begin{bmatrix} x_{k|k-1}^L \\ y_{k|k-1}^L \\ z_{k|k-1}^L \end{bmatrix} = \begin{bmatrix} x_{k-1|k-1}^L \\ y_{k-1|k-1}^L \\ z_{k-1|k-1}^L \end{bmatrix} + \begin{bmatrix} V_{x\,k-1|k-1}^L \\ V_{y\,k-1|k-1}^L \\ V_{z\,k-1|k-1}^L \end{bmatrix} T_s \tag{2}$$

where $Ts = 0.02\ s$ is the length of the discrete time step. For velocity prediction, the 3D acceleration measurements are integrated and transformed from the aircraft body-axis (B) to the local Cartesian navigation frame:

$$\begin{bmatrix} V_{x\,k|k-1}^L \\ V_{y\,k|k-1}^L \\ V_{z\,k|k-1}^L \end{bmatrix} = \begin{bmatrix} V_{x\,k-1|k-1}^L \\ V_{y\,k-1|k-1}^L \\ V_{z\,k-1|k-1}^L \end{bmatrix} + DCM(\phi_{k-1|k-1}^B, \theta_{k-1|k-1}^B, \psi_{k-1|k-1}^B) \begin{bmatrix} \tilde{a}_{x\,k}^B \\ \tilde{a}_{y\,k}^B \\ \tilde{a}_{z\,k}^B \end{bmatrix} T_s + \begin{bmatrix} 0 \\ 0 \\ g \end{bmatrix} T_s \tag{3}$$

where g is the earth's gravity, DCM stands for the Direction Cosine Matrix:

$$DCM(\phi,\theta,\psi) = \begin{bmatrix} c\psi\,c\theta & -s\psi\,c\phi + c\psi\,s\theta\,s\phi & s\psi\,s\phi + c\psi\,s\theta\,c\phi \\ s\psi\,c\theta & c\psi\,c\phi + s\psi\,s\theta\,s\phi & -c\psi\,s\phi + s\psi\,s\theta\,c\phi \\ -s\theta & c\theta\,s\phi & c\theta\,c\phi \end{bmatrix} \tag{4}$$

where 's' and 'c' are abbreviated sine and cosine functions respectively.

The aircraft Euler angles are predicted with the 3-axis angular rate measurements:

$$
\begin{bmatrix} \phi^B_{k|k-1} \\ \theta^B_{k|k-1} \\ \psi^B_{k|k-1} \end{bmatrix} = \begin{bmatrix} \phi^B_{k-1|k-1} \\ \theta^B_{k-1|k-1} \\ \psi^B_{k-1|k-1} \end{bmatrix} + \begin{bmatrix} \tilde{p}^B_k + \tilde{q}^B_k \sin\phi^B_{k-1|k-1} \tan\theta^B_{k-1|k-1} + \tilde{r}^B_k \cos\phi^B_{k-1|k-1} \tan\theta^B_{k-1|k-1} \\ \left(\tilde{q}^B_k \cos\phi^B_{k-1|k-1} - \tilde{r}^B_k \sin\phi^B_{k-1|k-1} \right) \\ \left((\tilde{q}^B_k \sin\phi^B_{k-1|k-1} + \tilde{r}^B_k \cos\phi^B_{k-1|k-1}) \sec\theta^B_{k-1|k-1} \right) \end{bmatrix} T_s \quad (5)
$$

The nine predicted state variables are then regulated by the GPS position and velocity measurements during the measurement update process with a simple observation equation:

$$
\mathbf{z}_k = \Big[\tilde{x}^L_k = x^L_k + v_x \quad \tilde{y}^L_k = y^L_k + v_y \quad \tilde{z}^L_k = z^L_k + v_z
$$
$$
\tilde{V}^L_{x\,k} = V^L_{x\,k} + v_{Vx} \quad \tilde{V}^L_{y\,k} = V^L_{y\,k} + v_{Vy} \quad \tilde{V}^L_{z\,k} = V^L_{z\,k} + v_{Vz} \Big]^T \quad (6)
$$

The solution of the GPS/INS sensor fusion problem follows the classis EKF approach as outlined in (Simon, 2006). The filter tuning is performed through the selection of the process noise covariance matrix Q and the measurement noise covariance matrix R. Specifically, the process noise is approximated by the sensor-level noise present on the IMU measurement.

$$
Q = diag([0,0,0,\sigma^2_{v_{ax}},\sigma^2_{v_{ay}},\sigma^2_{v_{az}},\sigma^2_{v_p},\sigma^2_{v_q},\sigma^2_{v_r}])T_s^2 \quad (7)
$$

where the first three zeros indicate that no uncertainty is associated with Equation (2). Similarly, the variance of the GPS measurement noise calculated with a ground static test is used for providing the R matrix:

$$
R = diag([\sigma^2_{v_x},\sigma^2_{v_y},\sigma^2_{v_z},\sigma^2_{v_{Vx}},\sigma^2_{v_{Vy}},\sigma^2_{v_{Vz}}]) \quad (8)
$$

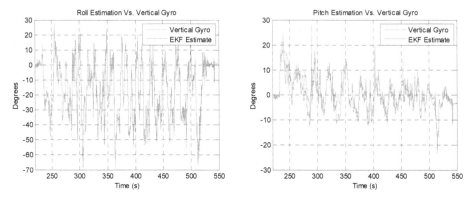

Fig. 8. Validation of the GPS/INS sensor fusion algorithm performance.

The performance and robustness of the attitude estimation algorithm was evaluated against multiple sets of flight data. Within these flights, a Goodrich VG34® mechanical vertical gyroscope was carried on-board to provide independent pitch and roll angle measurements and is used as the reference for evaluating the GPS/INS sensor fusion performance. The VG34 has a self-erection system, and reported accuracy of within 0.25° of true vertical. Figure 8 shows a comparison between the GPS/INS estimates and VG34 measurements on both roll and pitch channels for one of the May 27, 2011 flight tests. The mean absolute error and standard deviation error for roll estimation are 2.64° and 2.29° respectively in this particular flight. The mean absolute error and standard deviation error for pitch estimation are 2.22° and 1.93° respectively.

6. Avionic system testing

Extensive ground and flight testing experiments were performed to verify the functionality and performance of the Gen-V avionics design and to enable different aviation safety related flight experiments.

6.1 Avionics integration

The integration of avionics components into an airframe is constrained by many practical factors, such as aircraft balance, sensor alignment, signal interference, heat-dissipation, vibration damping, and user accessibility. Particularly, a key consideration for the avionic integration is to minimize the EMI effect. Within a sub-scale aircraft, the EMI issue is recurrent due to the close proximity of electrical components within a confined space. The effect of EMI includes reduced sensor measurement quality and disruptions of the command and control link, which could potentially lead to the loss of an aircraft. An integrated approach is used to mitigate the EMI problem. This include careful circuit design to reduce cross-interferences; providing redundancy on safety-critical components; proper shielding of main electronic components and cables; separation of EMI sources from R/C receivers; and reducing the number and length of cables. Once every avionics sub-system is installed, a comprehensive spectrum analysis and ground range tests are performed to identify residual EMI issues. Remaining problems can usually be alleviated through application of additional shielding materials, addition of ferrite chokes on selected cables, or through alternative antenna placements for RF modem and R/C receivers. Finally, a systematic ground range check procedure is performed before each flight to ensure a safe operation.

6.2 Ground testing

The ground testing procedure for the WVU Gen-V avionics system involves the following main categories:

1. *Hardware testing,* which includes the basic conductivity tests, evaluation of system power consumption and heat dissipation, EMI tests, and range tests for the R/C and data links;
2. *Software testing,* which includes the latency measurement of the real-time operating system and profiling the computational resource use by different software components;
3. *Hardware/software integration,* which includes the evaluation of sensor measurement quality, communication dropouts, PWM reading and generating accuracy, control system delay, and the functionality of the flight mode transition logics;

4. *Reliability testing*, which includes a number of duration tests under simulated dynamic operating environments;
5. *Calibration*, which includes the calibration of individual sensors, PWM reading and generating processes, individual control actuators, and pilot input devices such as R/C transmitter and the research pilot control station;
6. *Modelling*, this includes the development of mathematical models for the test-bed aircraft, actuators, propulsion systems, and sensors, as well as the identification of model parameters;
7. *Simulation*, which includes model-based simulation for initial validation of mission-specific research algorithms, and hardware-in-the-loop simulation for evaluating the integration between hardware and software sub-systems.

A flight test is considered after all related ground tests are performed.

6.3 Flight testing

Flight testing provides the final validation of the aircraft and its flight control system. However, it is also well known that experimental flight testing program, either with a full-scale or a sub-scale aircraft, is associated with substantial risks. A general strategy for flight risk mitigation focus on three steps:

1. *Prevent* the aircraft from entering an adverse flight condition;
2. Timely *identification* of the problem when an emergency situation develops;
3. *Recover* the aircraft or minimize its damage during the accident.

An adverse flight condition could be caused by improper/inadequate planning, pilot error, atmospheric condition, and aircraft sub-system (e.g. mechanical, electrical, power, control, and communication) failures. Quite often, an aviation accident has multiple inter-connected contributing factors (Boeing, 2009).

It is worth noting that the general objective of a fault-tolerant flight control research program with a sub-scale aircraft is usually to facilitate the development of the fault prevention, identification, and recovery methods for a full-scale manned aircraft. During flight experiments, the aircraft is often commanded to enter deliberately-planned adverse conditions, while minimizing other flight-associated potential risks. This high level of uncertainty, with both expected and unexpected failure contributing factors, provides valuable experiences and insights for understanding aviation accidents and the unique opportunity to practice and refine risk mitigation approaches.

6.3.1 Risk mitigation and flight testing protocol

Two effective approaches for improving the operational safety of a sub-scale flight testing program are incremental testing and the standardization of flight protocols. The incremental flight testing method utilizes a 'divide and conquer' approach to build-up individual sub-system capabilities and allows them to mature over a series of increasingly complex experiments. Each step should be a logic extension of previous steps, but should also be large enough to ensure a timely completion of the project. For example, an experiment to study the aircraft dynamics at high angle of attack flight conditions could be built upon the following key steps:

1. R/C flights for evaluating aircraft handling quality, stall characteristics, and payload capacity;
2. Data acquisition flights to evaluate avionics measurement quality and GPS/INS sensor fusion algorithm performance;
3. Closed-loop flights with a set of inner-loop control laws stabilizing the aircraft at the trim flight condition;
4. Closed-loop flights around the trim condition with OBES injection;
5. Closed-loop flights at high angle of attack conditions with OBES injection.

The standardization of flight testing protocols reduces human error both before and during the flight. It allows a systematic planning, resource allocation, testing, and inspection during the flight preparation. During the flight, having a standard procedure and flight pattern reduces pilot stress and improve the consistency among flights. Additionally, having an emergency handling procedure reduces the pilot reaction time and avoids making arbitrary decisions under adverse flight conditions. The flight testing protocol builds upon years of flight testing experience and provides a media to store and apply lessons learned from past mistakes.

A flow chart for the flight testing operation procedure developed at WVU is shown in Figure 9. A flight test session starts with a flight planning meeting in the lab discussing mission objectives, test methods, and personal responsibilities. A preliminary flight readiness review is normally performed a day before the flight date, following successful efforts in research algorithm development, ground test, and aircraft inspection.

At the airfield, another round of aircraft inspection and ground tests are performed to ensure that all aircraft sub-systems are operational after ground transportation. This is enforced with a flight preparation check-list, which covers airframe, avionics, R/C system, power system, firmware, research software, communication system, and the ground station. Additionally, the aircraft weight and balance are checked before the first flight of each aircraft. A final flight readiness review is then performed after the checklist is completed. Finally, a pre-flight pilot de-briefing discusses the flight procedures, research manoeuvres, and potential risks of this particular flight.

Once the aircraft is positioned at its starting position on the runway, the propulsion, R/C, and avionics systems are powered following an aircraft start-up procedure. A series of range tests are then performed to evaluate the R/C and data link range. A flight operation checklist is filled to verify the general functionality of the aircraft, such as control surface deflections, propulsion system condition, and R/C system fail-safe settings. A set of 'go/no-go' criteria, which includes wind-speed, wind-direction, communication range, and ground crew readiness, are then evaluated before a final approval of the flight by the flight director.

The flight operation itself follows a set of pre-defined take-off, trim, command hands-off, research, and landing procedures. In case of an emergency, such as a single engine failure, both engine failure, controller failure, actuator failure, aircraft upset condition, or changing weather condition, a set of specific emergency handling procedures are followed to abort the flight and recover the aircraft.

After landing and powering off the aircraft, flight data are downloaded and analysed in the field to provide an initial assessment of data quality and determine any potential issues. A post flight discussion session reviews the flight performance, problems encountered and pilot feedbacks. After returning to the lab, a detailed data analysis is performed, follows by a post-flight meeting to conclude the flight session.

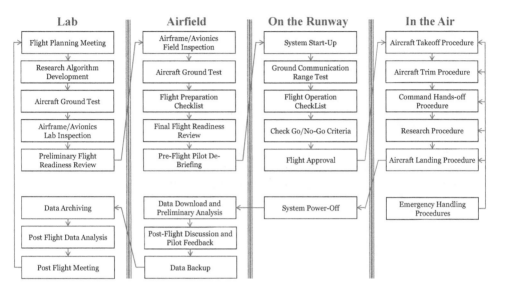

Fig. 9. WVU flight testing operation protocol.

6.3.2 Flight test examples

Two flight test examples are presented in this section to show the effectiveness of the designed avionics system. The first example is to collect data for identifying mathematical models of the 'Phastball' aircraft under high angle of attack flight conditions. The second example is to evaluate the human pilot performace with delayed control signals.

The objective of the first experiment is to study the aircraft dynamics under high angle of attack conditions. This is particularly important for T-tail aircraft, where the turbulent airflow from the stalled wing can blanket the elevators during a deep stall. For this experiment, the OBES manoeuver is designed with a multi-sine frequency-sweep approach (Klein & Morelli, 2006) to minimize disturbances to the flight condition. Specifically, it composes of six discrete frequency components ranging between 0.2 and 2.2 Hz. During the flight, a set of aircraft inner-loop controllers are activated with the *ctrl*-switch. The inner-loop controllers track zero degree roll angle and 12-degree pitch angle as reference inputs, while holding the throttle positions constant. After 2-seconds into the autonomous flight, a stream of 8-second of OBES manoeuvres are superimposed onto the elevator command generated by the inner-loop controllers. Several flight tests were performed with this configuration. Figure 10 shows a section of data collected from an October, 10, 2011 flight test.

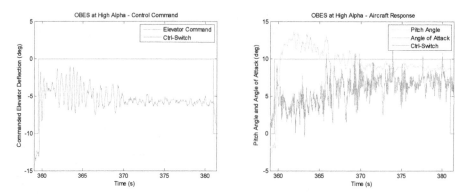

Fig. 10. Elevator control command (left) and aircraft response (right) with OBES manoeuvres at high angle of attack.

Within Figure 10, the red line indicates the turning on/off of the *ctrl*-switch. The angle of attack gradually increases to approximately 7 degrees with the deceleration of the aircraft. Additional flight experiments are planned to investigate higher angles of attack, as well as pre-stall and post-stall flight conditions.

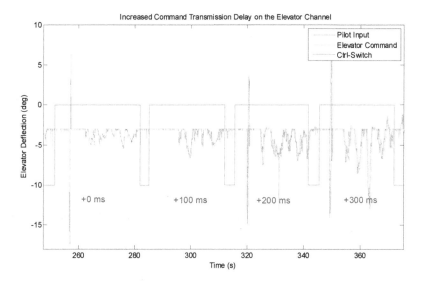

Fig. 11. Pilot elevator command vs. the actual elevator input during a command transmission delay experiment.

The objective of the second experiment is to study how the transmission delay of a fly-by-wire flight control system affects the handling quality of an aircraft. The on-board software is designed to relay the recorded pilot input to actuators with added delay. This occurs whenever the *ctrl*-switch was turned on, and a 100 ms increment is added to the total transmission delay during each *ctrl*-switch activation. The pilot flies the aircraft directly in the 'Manual Mode I' with the ctrl-switch off. During the flight test, the pilot turns on the ctrl-switch at the beginning of a straight path. The pilot first injects an elevator doublet, wait until it settles, and then performs a turn manoeuvre. This process repeats multiple times in flight but with increasing transmission delay up to 300ms.

The first experiment of this kind was performed on October 18, 2011. The pilot reported "...the whole flight was normal, and decay in elevator was negligible..." during the post-flight discussion. However, flight data collected clearly indicates increased pilot activity with increased command transmission delay, which is shown in Figure 11. A later experiment with a different pilot showed similar results. Additional flight experiments are planned to investigate larger transmission delay, random data dropouts, and flight conditions that require precise control actions.

7. Conclusions

The use of sub-scale research aircraft provides unique opportunities for investigating adverse flight conditions that are too risky or costly to be tested on a full-scale aircraft. It can be considered as an intermediate validation tool between a flight simulator and a full-scale aircraft. It allows the testing of different system design, modelling, control, fault detection, and risk mitigation approaches within a realistic physical environment.

Sub-scale flight testing for fault tolerant flight control research also poses many challenges to the avionic system design: 1) it requires new capabilities for simulating different aircraft upset and failure conditions; and 2) it requires a flexible interface for integrating both human and machine decision-making capabilities; and 3) it needs to be reliable and fault-tolerant to both planned and unexpected failures. The Gen-V avionics system, designed and being developed at WVU meets these complex research requirements, along with strict power, weight, size, and cost limitations. Preliminary flight testing results demonstrate the capability of the proposed avionics design and its flexibility in supporting a variety of research objectives.

8. Acknowledgment

This study was conducted with partial support from NASA grant # NNX07AT53A and grant # NNX10AI14G

9. Appendix A: List of achronoymes

ADC – Analog to Digital Conversion
CAS – Control Augmentation System
CCSS – Confirmed Ctrl Switch Signal
COTS – Commercial-off-the-Shelf
EKF – Extended Kalman Filter
EMI – Electromagnetic Interference
EMP – Embedded Micro-Processor
FCS – Flight Control System
FES – Failure Emulation Software
GCS – Ground Control Station

GNC – Guidance, Navigation, Control
GPC – General-Purpose Computer
GPS – Global Positioning System
GPT – General-Purpose Timer
HUD – Heads-Up Display
IMU – Inertial Measurement Unit
INS – Inertial Navigation System
OBES – On-Board Excitation System
OS – Operating System
PCC – Pilot Control Command
PID – Parameter Identification
PPM – Pulse-Position Modulation
PWM – Pulse-Width Modulation
R/C – Remote Control
RF – Radio Frequency
RTAI – Real-Time Application Interface
SAS – Stability Augmentation System
SCC – Software-generated Control Commands
SCI – Serial Communication Interfaces
SPI – Serial Peripheral Interface
SPOF – Single Point of Failure
UAV – Unmanned Aerial Vehicle
WVU – West Virginia University

10. References

Ambrosia, V.G.; Brass, J.A.; Greenfield, P. & Wegener, S. (2004). *Collaborative Efforts in R&D and Applications of Imaging Wildfires*, US Forest Service. Available: http://geo.arc.nasa.gov/sge/WRAP/projects/docs/RS2004_PAPER.PDF.

Boeing Commercial Airplanes, (2009) Statistical Summary of Commercial Jet Airplane Accidents, World Wide Operations, 1959-2008, Seattle, WA, Available: http://www.boeing.com/news/techissues.

Chao, H.Y.; Jensen, A.M.; Han, Y.; Chen, Y.Q. & McKee, M. (2009). AggieAir: Towards Low-cost Cooperative Multispectral Remote Sensing Using Small Unmanned Aircraft Systems, Chapter, *Advances in Geoscience and Remote Sensing*, Gary Jedlovec, Ed.Vukovar,Croatia:IN-TECH,pp.467–490.

Christophersen, H.B.; Pickell, W.J.; Koller, A.A.; Kannan, S.K & Johnson, E.N. (2004). Small Adaptive Flight Control Systems for UAVs using FPGA/DSP Technology, *Proceedings of the AIAA "Unmanned Unlimited" Technical Conference, Workshop, and Exhibit*, Chicago, IL, September, 2004.

Cione, J.J.; Uhlhorn, E. W.; Cascella, G.; Majumdar, S. J.; Sisko, C.; Carrasco, N.; Powell, M. D.; Bale, P.; Holland, G.; Turlington, P.; Fowler, D.; Landsea, C. W. & Yuhas, C. L. (2008). The First Successful Unmanned Aerial System (UAS) Mission into a Tropical Cyclone (Ophelia 2005), *12th Conference on IOAS-AOLS*, New Orleans, LA, January 2008.

Evans, J.; Inalhan, G.; Jang J.S.; Teo, R. & Tomlin, C.J. (2001). DragonFly: a Versatile UAV Platform for the Advancement of Aircraft Navigation and Control, *Digital Avionics*

Systems, DASC. The 20th Conference, vol.1, pp.1C3/1-1C3/12, Daytona Beach, FL, October, 2001.

Griffiths, S.; Saunders, J.; Curtis, A.; Barber, B.; McLain, T. & Beard, R. (2006). Maximizing Miniature Aerial Vehicles, *IEEE Robotics & Automation Magazine*, vol.13, no.3, pp. 34-43, Sept. 2006.

Gross, J.; Gu, Y.; Rhudy, M.; Gururajan, S. & Napolitano, M.R. (2011). Flight Test Evaluation of Sensor Fusion Algorithms for Attitude Estimation, *IEEE Transactions on Aerospace and Electronic Systems*, In Press, June, 2011.

Gu, Y.; Campa, G.; Seanor, B.; Gururajan, S. & Napolitano, M.R. (2009). Autonomous Formation Flight – Design and Experiments, Chapter, *Aerial Vehicles*, ISBN 978-953-7619-41-1, I-Tech Education and Publishing, Austria, EU, Chapter 12, pp. 233-256.

How, J.P.; Bethke, B.; Frank, A.; Dale, D. & Vian, J. (2008). Real-Time Indoor Autonomous Vehicle Test Environment, *IEEE Control Systems Magzine*, vol.28, no.2, pp.51-64, April, 2008.

Jordan, T. L.; Foster, J. V.; Bailey, R. M.; & Belcastro, C. M. (2006). AirSTAR: A UAV Platform for Flight Dynamics and Control System Testing, *25th AIAA Aerodynamic Measurement Technology and Ground Testing Conference*, San Francisco, CA, June, 2006.

Jourdan, D.B.; Piedmonte,M.D.; Gavrilets,V. & Vos,D.W. (2010). Enhancing UAV Survivability Through Damage Tolerant Control, *AIAA Guidnace Navigation and Control Conference*, Toronto, Ontario, Canada, August, 2010.

Klein, V. & Morelli, E.A. (2006). *Aircraft System Identification – Theory and Practice*, AIAA Education Series, AIAA, Reston, VA.

Liebeck, R.H. (2004). Design of the Blended Wing Body Subsonic Transport, *Journal of Aircraft*, pp. 10-25, Vol. 41, No. 1, January–February, 2004.

Miller, J.A.; Minear, P.D.; Niessner, A.F.; DeLullo, A.M.; Geiger, B.R.; Long, L.L. & Horn, J.F. (2005). Intelligent Unmanned Air Vehicle Flight Systems, *Infotec@AIAA*, Arlington, Virginia, September, 2005.

Murch, A. M. (2008). A Flight Control System Architecture for the NASA AirSTAR Flight Test Infrastructure, *AIAA Guidance, Navigation, and Control Conference*, Honolulu, HI, August, 2008.

NRC (National Research Council) (1997). Aviation Safety And Pilot Control: Under-standing and Preventing Unfavorable Pilot-Vehicle Interactions, *National Academy Press*.

Perhinschi, M.; Napolitano, M.R.; Campa, G.; Seanor, B.; Gururajan, S. & Gu, Y. (2005). Design and Flight Testing of Intelligent Flight Control Laws for the WVU YF-22 Model Aircraft, *AIAA Guidance, Navigation, and Control Conference*, San Francisco, California, August, 2005.

Phillips, K.; Gururajan, S.; Campa, G.; Seanor, B.; Gu, Y. & Napolitano, M.R. (2010). Nonlinear Aircraft Model Identification and Validation for a Fault-Tolerant Flight Control System, *AIAA Atmospheric Flight Mechanics Conference*, Toronto, Ontario, Canada, August 2010.

Planecrashinfo.com (2011). Causes of Fatal Accidents by Decade, Avaliable: http://planecrashinfo.com/cause.htm.

Rhudy, M.; Gu, Y.; Gross, J. & Napolitano, M. R. (2011). Sensitivity Analysis of EKF and UKF in GPS/INS Sensor Fusion, *AIAA Guidance, Navigation, and Control Conference*, Portland, OR, August, 2011.

Simon, D. (2006). *Optimal State Estimation: Kalman, H-Innity, and Nonlinear Approaches*. Wiley & Sons, 1. edition.

Yeh, Y.C. (1998). Design Considerations in Boeing 777 Fly-By-Wire Computers, *Third IEEE International High-Assurance Systems Engineering Symposium*, Washington, DC, November, 1998.

9

Study of Effects
of Lightning Strikes to an Aircraft

N.I. Petrov[1], A. Haddad[2], G.N. Petrova[1], H. Griffiths[2] and R.T. Waters[2]

[1]*Istra, Moscow region,*
[2]*Cardiff University,*
[1]*Russia*
[2]*United Kingdom*

1. Introduction

It is difficult to avoid thunderstorm regions by aircraft, so that on average every commercial airliner is struck by lightning once per year. Defining test and design criteria of aircraft is becoming important since aircraft safety is increasingly dependent on electronic equipment and the development of new materials (carbon composites, etc.) to replace the metallic airframes.

In-flight statistics show that most strikes occurred 3-5 km above sea level, where the temperature is $\sim 0°C$ (Uman & Rakov, 2003; Larsson, 2002). There are two different types of lightning strikes to aircraft. The first type is that the aircraft initiates the lightning discharge when it is found in the intense electric field region of a thundercloud, and the second is the interception by the aircraft of an approaching lightning leader. The mechanism for lightning initiation by aircraft is often explained using the "bidirectional leader" theory (Clifford & Casemir, 1982; Mazur, 1989; Mazur et al., 1990; Mazur & Moreau, 1992), which describes the aircraft-initiated lightning process as a positive leader starting from the aircraft in the direction of the ambient electric field; this is followed, a few milliseconds later, by a negative leader developing in the opposite direction. This order of events is a consequence of the lower electric strength of air in the vicinity of a divergent (anode) field. The ambient thundercloud electric field measured under such conditions is typically in the range 50 - 100 *kV/m* (Marshall & Rust, 1991).

Radome "measles" (coloured spots on the inner radome surface) have been observed in many instances during service (Lalande et al., 1999; Ulmann et al., 1999). Each spot corresponds to a pin hole through the sandwich panel of the radome material. A possible explanation of the origin of these pin holes is that they were caused by breakdown due to double-layer charge accumulation on the radome. However, the physical mechanisms of the occurrence of "measles" are not fully established yet.

The purpose of this chapter was to investigate the physical processes involved in lightning strikes to aircraft and to compare simulation results with other studies involving instrumented aircraft flying in thunderstorms. 3-*D* electric field calculations were performed to determine the field distributions at the nose of aircraft and inside the dielectric radome (nosecone). The influence of the thickness and dielectric constant of the radome wall on the electric field penetration inside the radome was also investigated. The screening effect caused by ice and water layers on the radome wall is demonstrated. A new proposal for radome protection is made possible by the development of strips using materials such as non-linear ZnO, which behave as dielectrics under low-field conditions and acquire properties of conductors if the external electric field exceeds the critical value. Experimental tests of the strips on a real aircraft radome were carried out, and the test results reported in this paper.

2. Lightning attachment to aircraft

It was recently reported that about 90% of lightning strikes to aircraft are initiated by the aircraft (Uman & Rakov, 2003). This indicates that the aircraft extremities provide the region of high electric field needed to initiate a lightning discharge by enhancing the ambient electric field. The aircraft geometry and ambient atmospheric conditions are the most important factors in determining the local electric field intensification. Since pressure, absolute humidity and temperature decrease with increasing altitude, the variation of streamer properties with altitude can be inferred from laboratory experiments and incorporated into lightning modelling.

It is inferred from (Petrov & Waters, 1994, 1995) that the electric field needed to initiate a lightning discharge at 4km altitude is only about half of the value at sea level. Calculations show that the required striking distance increases significantly with increasing altitude, causing a corresponding increase in the risk of lightning strikes for aircraft in flight. It is shown, in the following, that ambient electric fields of between 50-80 kV/m can initiate positive leaders at the nose of aircraft at such altitudes.

2.1 Aircraft-initiated lightning

Consider the aircraft body as an electrically floating conducting ellipsoid placed in a uniform ambient electric field E_0 (Fig.1). An analytical expression may be obtained for the enhanced electric field in the vicinity of the nose for the case where the major axis is parallel to E_0 (Petrov & Waters, 1994):

$$E(x,a,b) = E_0 \left\{ 1 - \frac{ar\tanh(aA^{1/2}/x) - aA^{1/2}/x}{ar\tanh A^{1/2} - A^{1/2}} + \frac{A}{(x^2/a^2 + b^2/a^2 - 1)} \frac{aA^{1/2}/x}{(ar\tanh A^{1/2} - A^{1/2})} \right\} \tag{1}$$

where $A = 1 - b^2/a^2$, a and b are the half-length and half-width of the ellipsoid and $(x - a)$ is the distance from the ellipsoid tip.

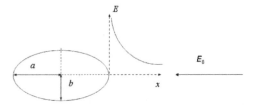

Fig. 1. Aircraft model representation and field intensification.

For given ellipsoid parameters, it is possible to determine the critical value of the ambient electric field which predicts a successful leader development from the aircraft. Using the criteria from (Petrov & Waters, 1994), we find that ambient field magnitudes of $E_{cr} \approx 50$ - 80kV/m (Fig. 2). This is insufficient at sea level to initiate leaders from the aircraft tip. However, at an altitude of 4000m, where the relative air density is around 0.58, triggering of leaders originating from the nose could certainly occur. Ambient fields of 50kV/m agree well with the fields measured inside storm-cloud, consistent with the in-flight measurements of lightning strikes to aircraft (Lalande et al., 1999).

The critical electric field dependence on the half-length of the aircraft, can be approximated with high accuracy using the empirical relationship

$$E_{cr} \cong 570 \cdot a^{-0.68} , \qquad (2)$$

where a is in m, and E_{cr} in kV/m.

Similar relationship with slightly different coefficient was obtained in (Petrov & D'Alessandro, 2002) for earthed structures.

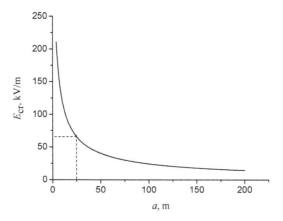

Fig. 2. Critical ambient electric field as a function of aircraft half-length (E_{cr} = 65 kV/m at a=25 m, b=3m.).

2.2 Aircraft-intercepted lightning

An aircraft can, in principle, intercept an approaching lightning leader, although no direct evidence is available. Nevertheless, in this case, the striking distance concept usually used for earthed structures may be applied to estimate the risk factor. The striking distance and the probability of lightning strikes are functions of aircraft geometry and lightning current. Electric field intensification of the field of a nearby lightning leader as a function of the distance from the aircraft tip is presented for different values of lightning peak current in Fig. 3. The aircraft is again modelled as an ellipsoid with half-width of 3m and half-length of 25m. The lightning leader channel is modeled by a charge per length, q, and leader tip charge, Q, at a distance, S, from the aircraft. The values for q and Q correspond to a prospective lightning return stroke current i_0, evaluated from (Petrov & Waters, 1995), i.e.

$$q \approx 0.43 \cdot 10^{-6} i_0^{2/3} \quad [\text{C/m, A}], \tag{3}$$

Note that there are similar relationships between the leader channel charge and the return stroke current obtained from other models. A review of data concerning this relationship was made in (Cooray et al., 2004).

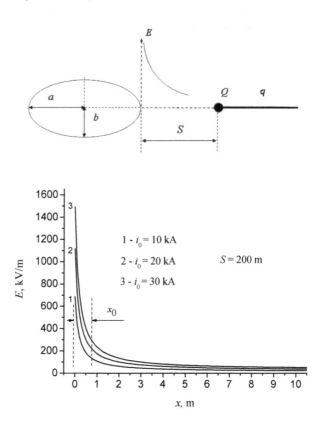

Fig. 3. Electric field intensification as a function of distance from the aircraft tip for different values of lightning peak current.

In Fig. 4, the striking distance of negative lightning to the aircraft as a function of lightning current is presented for different altitudes above sea level. Note, that for positive lightning, these distances are substantially less than those obtained for negative polarity lightning (Petrov & Waters, 1999).

A semi-quantitative estimate of the risk of lightning strike interception by an aircraft can be obtained from the concept of attractive area as used in lightning protection standards for ground structures, which can also be derived from lightning models (Petrov & Waters, 1995). For a grounded structure of the size of a commercial aircraft, the attractive area to a powerful lightning stroke of 100kA is of the order of 0.2km². At 4000m altitude, this would increase to 0.6km². Then, if the flash activity (cloud-cloud and cloud-ground) is N flashes/km²/s, the aircraft would be expected to intercept 0.6N flashes/s. Active storms can generate 2 flashes/minute over 10 km², which suggests an interception rate of 1 per 500s at the heart of a storm.

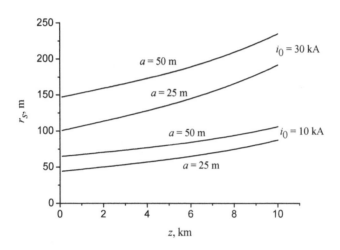

Fig. 4. Lightning interception distances by aircraft of different half-lengths as function of altitude above sea level for lightning peak current values of 10 kA and 30 kA.

3. Electric field around radomes

Radar and communications antennae are usually located at the nose or tail of the aircraft where lightning is most likely to attach. Lightning strikes damage non-metallic radomes, so the diverter strips were developed to mitigate this problem. The diverter strips screen the lightning induced electric fields on the antenna surface, i.e. they move the internal streamer initiation points forward so that strips cause the collapse of electric field inside the radome. Solid strips (permanent conductors) have been used for this purpose. However, they were found to interfere with antenna radiation patterns because they usually extend beyond the antenna. For this reason, segmented diverter strips were developed to reduce the interference effects on antenna radiation (Amason et al., 1975; Plumer & Hoots, 1978). Although they have better electromagnetic transparency for radar, segmented strips need a significant voltage gradient to light up, and their efficiency needs to be further proved.

3.1 Electric field distribution at radome without strips

For a simplified analytical calculation of 3-D electric field, consider a hemi-spherical radome with thickness d placed in uniform field E_0 (Fig. 5). This is equivalent to the floating dielectric hollow sphere (permittivity ε, internal and external radii a and b) placed in the electric field E_0.

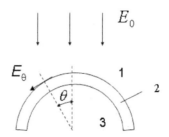

Fig. 5. Simplified model of radome exposed to electric field E_0.

The analytical solution of Laplace's equation for the potentials outside the sphere (Region 1) and inside the sphere (Region 3) can be obtained as:

$$\varphi_1 = -E_0 \cos\theta \left(r - \frac{A}{r^2} \right), \quad [r > b];$$

$$\varphi_3 = -BE_0 r \cos\theta, \qquad [r < a]$$

(4)

and the potential inside the dielectric layer (Region 2)

$$\varphi_2 = -CE_0 \cos\theta \left(r - \frac{D}{r^2} \right), \quad [a < r < b]$$

(5)

where A, B, C, D are constants determined from the continuity condition for φ and $\varepsilon \partial\varphi/\partial r$ on the boundaries of regions 1-2 and 2-3. Calculation of these constants leads to the following expressions:

$$A = a^3 \left\{ 1 - \frac{3\left[1 + 2\varepsilon + (\varepsilon - 1)b^3 / a^3 \right]}{(\varepsilon + 2)(2\varepsilon + 1) - 2(\varepsilon - 1)^2 b^3 / a^3} \right\},$$

$$B = \frac{9\varepsilon}{(\varepsilon + 2)(2\varepsilon + 1) - 2(\varepsilon - 1)^2 b^3 / a^3},$$

$$C = \frac{3(2\varepsilon + 1)}{(\varepsilon + 2)(2\varepsilon + 1) - 2(\varepsilon - 1)^2 b^3 / a^3}, \quad D = -\frac{b^3(\varepsilon - 1)}{2\varepsilon + 1}.$$

(6)

For the radial and tangential components of the electric field outside the radome surface, we obtain

$$E_r = -\frac{\partial\varphi}{\partial r} = E_0 \cos\theta \left(1 + \frac{2A}{r^3} \right), \quad [r > b]$$

(7)

$$E_\theta = -\frac{1}{r}\frac{\partial\varphi}{\partial\theta} = -E_0 \sin\theta \left(1 - \frac{A}{r^2} \right), \quad [r > b]$$

In Figs. 6a and 6b, the radial and tangential electric field distributions are presented for radomes having different dielectric constants. It is seen that the screening of the electric field by the radome itself increases when the dielectric constant of the radome material increases.

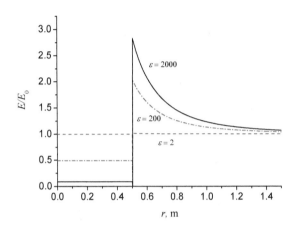

a.   Radial electric field distribution inside and outside the one-layer semi-spherical radome

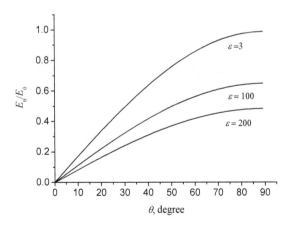

b.   Tangential electric field distribution inside the one-layer semi-spherical radome

Fig. 6. Electric field distribution in the vicinity of a radome.

i.   2-layer radome wall

By analogy, the potentials and electric fields may be obtained for the 2-layer radome wall placed in the field E_0:

$$\varphi_1 = -E_0 \cos\theta\left(r - \frac{A}{r^2}\right), \quad [r > c]$$

$$\varphi_2 = -CE_0 \cos\theta\left(r - \frac{D}{r^2}\right), \quad [a < r < c]$$

$$\varphi_3 = -FE_0 \cos\theta\left(r - \frac{G}{r^2}\right), \quad [b < r < a];$$

$$\varphi_4 = -BE_0 r \cos\theta, \quad [r < b] \tag{8}$$

$$A = c^3 - C(c^3 - D), \quad B = \frac{F(b^3 - G)}{b^3}$$

$$C = \frac{3\varepsilon_1 c^3}{2\varepsilon_1(c^3 - D) + \varepsilon_2(c^3 + 2D)},$$

$$D = \frac{a^3\left[1 - (\varepsilon_2 / \varepsilon_3)(a^3 - G)/(a^3 + 2G)\right]}{1 + 2(\varepsilon_2 / \varepsilon_3)(a^3 - G)/(a^3 + 2G)},$$

$$F = \frac{C(a^3 - D)}{a^3 - G}, \quad G = \frac{b^3(1 - \varepsilon_3 / \varepsilon_4)}{1 + 2\varepsilon_3 / \varepsilon_4},$$

where $b < a < c$, with b the internal radius of the inner layer, a and c are the internal and external radii of the exterior layer, ε_1, ε_2, ε_3, ε_4 are the dielectric constants of outside medium (air), exterior and interior layers, and inside medium (air), accordingly.

Radial and tangential components of the electric field outside the radome surface are expressed by

$$E_r = -\frac{\partial\varphi}{\partial r} = E_0 \cos\theta\left(1 + \frac{2A}{r^3}\right), \quad [r > c]$$

$$E_\theta = -\frac{1}{r}\frac{\partial\varphi}{\partial\theta} = -E_0 \sin\theta\left(1 - \frac{A}{r^2}\right), \quad [r > c] \tag{9}$$

In Fig. 7, the electric field distributions inside and outside the two-layer semi-sphere radome are presented for different values of dielectric constants of layers. It can be seen that the field intensification at the tip of a radome increases with the dielectric constant of the radome layers.

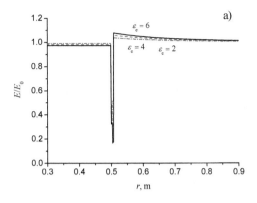

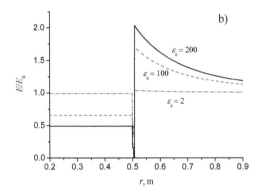

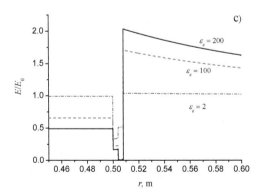

Fig. 7. Electric field distribution inside and outside the two-layer semi-sphere radome: *a*) $\varepsilon_i=3$ for internal layer; $\varepsilon_{e1} = 2$, $\varepsilon_{e2} = 4$, $\varepsilon_{e3} = 6$ for external layer; *b*) $\varepsilon_i = 3$; $\varepsilon_{e1} = 2$, $\varepsilon_{e2} = 100$, $\varepsilon_{e3}=200$; c) expanded scale of *b*).

ii.   Effect of ice and water layers

In in-flight environmental conditions, the radome may be covered by ice or water layers. The tests on radomes in rain and icing conditions were conducted recently (Hardwick et al., 1999, 2003), and it was shown that the ice layers increase the light up voltages by a factor 2 to 3.

Calculations of the electric field distributions in the case of ice and water on the radome surface show that radome produces significant shielding effect (Fig. 8). In this case, the lightning leader can be initiated from the radome tip, so the strips will not operate as usual. In Fig. 8, the electric field distributions are presented for different values of permittivity of the radome wall material. The radar is represented by a conducting hemisphere having a radius of 0.2m. Note, that for a wide range of frequencies, the dielectric constants of water and ice are equal to $\varepsilon_{H2O} = 87.9$ and $\varepsilon_{ice} = 99$, respectively (Handbook of Chemistry, 2001).

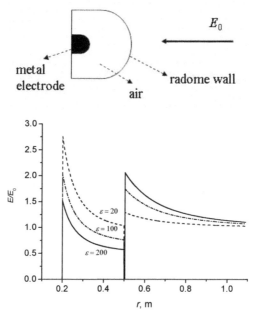

Fig. 8. Electric field distribution inside and outside the hemisphere radome with different dielectric constants of radome material.

3.2 Electric field shielding effect of strips

As was shown above, the electric field inside a dielectric radome is not disturbed significantly by the radome wall itself, so the radome does not produce screening effects. Low-level shielding permits the inception of a discharge from the internal electrode, so the solid strips are usually used to produce the shielding effect. However, high quality shielding has undesirable interference effects on antenna radiation. Therefore, the optimal length and number of strips should be determined. In the following, we consider a conical shaped radome with a base diameter of 0.7m (Fig. 9). For this radome, electric field measurement results at its base were reported (Ulmann et al., 2001; Delannoy et al., 2001), which allows comparison of simulations with experimental data. Here, solid strips were considered as inclined isolated rods in a uniform external electric field, since the analytical expressions for the electric field distribution exist in this case. In Fig. 10, the electric field at the radome base is shown as a function of strip length for different numbers of strips. It can be observed that the electric field at the radome base decreases by 50 % if 6 solid diverter strips of 0.4m length were installed on the radome surface. This is in good agreement with the measurements (Ulmann et al., 2001).

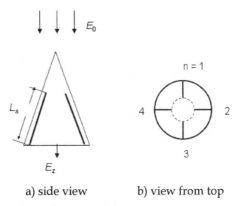

a) side view   b) view from top

Fig. 9. A conical radome with conducting solid strips:

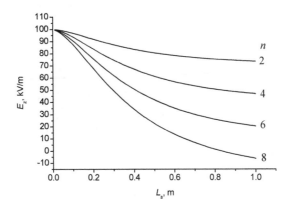

Fig. 10. Calculated electric field at the radome base as a function of strip length for different numbers of strips

4. Laboratory lightning impulse tests

Preliminary tests on a radome, used on a commercial aircraft, with a thickness of ~5 mm, a diameter of ~ 1.6 m and having six solid strips of 1m length were performed in the high voltage laboratory at Cardiff University. Lightning impulses of 1.2/50 shape, positive and negative polarity, were applied to the output electrode (sphere of 10*cm* diameter or rod with spherical end of 1.2*cm* diameter), which was placed at different distances (10-30 *cm)* from the surface of the radome. Breakdown channels were recorded using a video-camera having a picture rate of 50 fps.

4.1 Segmented diverter strips

Tests were also conducted on two commercially available segmented diverter strips of one meter length each. The diverter strips were attached to the aircraft radome surface for testing. It was found that the diverter with smaller buttons (segment diameter 1.524 mm) has higher breakdown voltage.

The segmented diverters had breakdown voltages of 50-60*kV* while the time to breakdown t_{br} varied between 3 and 7μs. This corresponds to leader velocities v_l in the range 15 to 30 *cm/μs*, which is ten times higher than usually registered leader velocities in long air gaps. Dependence of the applied voltage on the polarity is weak, if the rod-type high voltage electrode is placed close (~10-15cm) to the strip end.

Although the segmented strips have good diversion properties, tests have shown problems with multi-impulse lightning strikes. After a number of strikes, damage was observed on the strip buttons (Fig. 11). However, the resistance of the strips after tests was still more than 600 MΩ. This indicates that the discharge current mainly flows not through the strip buttons but in air over the strip.

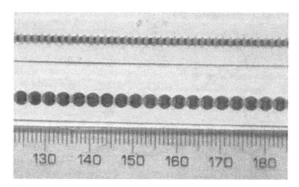

Fig. 11. Segmented diverter strips after the test.

4.2 Isolated multiple-electrode diverter

Isolated rings or disks with diameters of 17*mm* and 20*mm* with a separation of 5-15*cm* were mounted on the radome surface with the help of dielectric tape. These types of strips have several advantages: (a) they have negligible interference effects on antenna radiation due to the small total surface of metal elements and (b) they do not initiate a leader discharge before an approaching lightning leader streamer zone attaches to the radome surface. Both positive and negative polarity impulses of amplitude 200-250 kV were applied to gaps of 10-20cm between the electrode and the radome tip (Fig. 12 and Fig. 13).

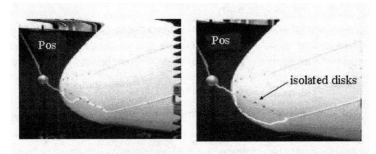

Fig. 12. Influence of isolated disks on the trajectory of breakdown under positive impulse

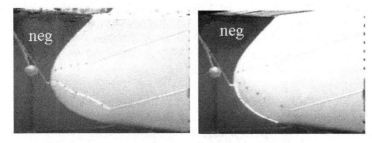

Fig. 13. Influence of isolated disks on the trajectory of breakdown under negative impulse

Breakdown occurs between the electrode and the closest point on the radome surface, propagating further along the radome surface until the end of the solid strip, even if the air gap distance between the electrode and the end of the solid strip is shorter. This indicates that the breakdown voltage along the surface of the radome is lower than the breakdown voltage in an air gap. The time to the surface breakdown was t_{br}~ 30-50 μs depending on the distance between the electrode and the end of the solid strip on the radome wall. This corresponds to a leader velocity v_l ~ 2 $cm/\mu s$, which is usually recorded in long air gaps. In the case of the isolated multiple-electrode diverters, the time to breakdown decreases by a factor of 3-4, i.e. the leader velocity becomes v_l ~ 6-8 $cm/\mu s$. The decrease of the breakdown time indicates that simultaneous development of the discharge in the gaps between different isolated electrodes.

The light up electric field was about 3.3 kV/cm, which is close to typical light up voltages with D waveform for the segmented strips (Hardwick et al., 1999).

Tests have shown that the isolated electrode strips divert the discharge channel of both polarities. Leaders develop along the surface without any damage to it. For the same applied voltage, the breakdown gap with the isolated multiple-electrode diverter strip can be twice as long as the gap without strip. The diversion ability of isolated multiple-electrode diverter strip is higher for negative polarity discharge than for positive discharge (Figs. 12, 13). This is due to different mechanisms of breakdown for negative and positive polarity discharges (Petrov & Waters, 1999).

4.3 Flashover across the radome wall

The discharge develops along the surface of the radome even if the air gap distance between the output electrode and the termination of the strips is shorter. This indicates that the breakdown voltage along the dielectric surface is lower than the breakdown voltage in air.

A leader discharge can be initiated from the internal radar antenna. In the model, the antenna was represented by a grounded metal hemisphere at the radome base. The leader channel from the antenna was modeled as a metal rod of different lengths connected with the antenna.

The laboratory experiments have shown that both positive and negative polarity discharges can cause a puncture through the radome wall when the internal electrode (antenna) extends beyond the strips and, hence, when it is no longer screened.

In Fig. 14, the flashover path can be seen initially propagating along the surface and then passing through the radome wall to the internal grounded electrode. The distance from the surface puncture point to the grounded outer electrode was only 7.5cm. This indicates that a voltage drop of less than 20 kV is sufficient to cause a flashover across the radome wall.

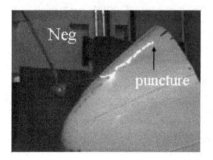 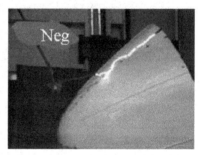

Fig. 14. Puncture through the radome wall with an earth electrode inside the radome.

4.4 Diverter strips with ZnO material

Segmented strips consisting of ZnO material between Al segments of 3x3 mm size were designed and tested (Fig. 15). Experiments have shown that the influence of ZnO material on the discharge properties of strips depends on the distances between the segments. Although no significant influence was observed for gaps $d > 10$ mm, at $d \sim 1$-3 mm, the influence of ZnO material becomes significant. The competitive breakdown tests showed that all discharges pass through the strip consisting of ZnO material, which indicates that electric fields created between the segments are sufficient for the ZnO material to become conductive. The breakdown time for these strips is comparable to that of commercial segmented strips. The velocity of leader propagation increases 4-5 times in comparison to the velocity of the surface leader discharge without the strips.

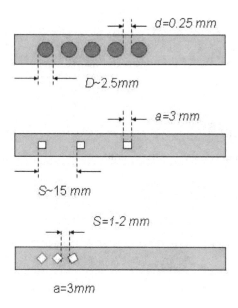

Fig. 15. Designed diverter strips with ZnO material.

5. 3D numerical computation of electric field around radomes

The electric field and potential distributions inside and outside the aircraft radome placed in an external electric field were analyzed using COULOMB software which is based on the boundary element method. This results of this analysis were used to determine the necessary number and length of strips to be utilized to provide the radome with the optimised lightning protection using strips.

A simulation model of the aircraft radome having a hemispherical shape placed in a uniform ambient electric field was used in a plane-plane gap (Fig. 16). The gap length is 5.2 m and the applied voltage is 2 MV. The dielectric hemispherical radome is placed on top of a metal cylinder of 1.5 m length to simulate the end of the fuselage. The hemispherical radome has a radius of 0.5m and a thickness of 4mm and a dielectric constant ε_r = 10. Solid strips of 1cm width and 3mm thickness were considered. The segmented strips have a 5mm diameter and a 3mm thickness of and a gap distance of 1mm. The distance between the radome tip and the upper electrode is 2m. The distance between the bottom of the cylinder and the bottom electrode is 1.2m.

Fig. 17 shows the solid and segmented strips attached to the radome surface. Fig. 18 shows examples of computed voltage contours.

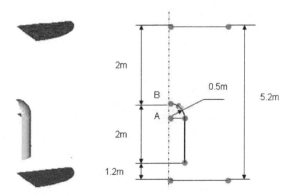

Fig. 16. Model representation: semi-spherical radome.

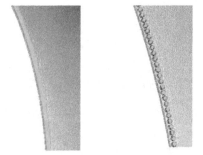

Fig. 17. Modeled solid (thickness: 3mm, width: 10 mm) and segmented (thickness: 3mm, radius: 2.5 mm, gap: 1 mm) strips.

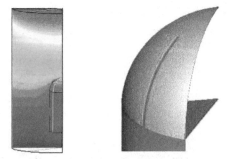

Fig. 18. Voltage contour and section of a radome with a solid strip.

Tables 1 and 2 summarise the computed magnitudes of electric field at the radome base and tip. It can be observed that the shielding effect increases with the length of solid strips and the number of strips. The electric field at the base of the radome is only 50% of the external field if 6 solid strips of 0.5m length are used. Segmented strips do not produce any visible shielding effects.

Detailed analaysis has shown that an increase of the number of solid strips results in a decrease in the electric field at the base of the radome. On the other hand, the electric field was forced out to the frontal area of the radome, so that too strong shielding of the internal electrode can cause undesired field intensification at a radome front. This is a disadvantage with solid strips, in addition to their interference effect on the radiation field from the antenna. In the case of segmented strips, there is no shielding effect. This indicates that there will be no interference effect with the radiation field until the breakdown along the strip takes place, under which condition the strip behaves like a conductor.

	Solid strips					
Number of strips	4		6		8	
Length, m	0.25	0.5	0.25	0.5	0.25	0.5
$E(A)$, kV/m	482	335	456	270	435	226
$E(B)$, kV/m	493	534	515	588	524	570

Table 1. Electric field magnitudes at radome base (point A in Fig.16) and radome tip (point B in Fig. 16) for different numbers of solid strips.

	Segmented strips					
Nunber of strips	4		6		8	
Length, m	0.25	0.5	0.25	0.5	0.25	0.5
$E(A)$, kV/m	517	524	526	508	551	515
$E(B)$, kV/m	477	472	481	497	477	475

Table 2. Electric field magnitudes at radome base (point A in Fig. 16) and radome tip (point B in Fig. 16) for different numbers of segmented strips.

6. Discussion

The radome simulations described in this chapter show clearly that the critical electric field magnitude, which is necessary to originate leaders from the aircraft tip, decreases with the aircraft length. The magnitude of the critical electric field decreases from 100 kV/m to 40 kV/m as the aircraft length increases from 20m to 100m. These values are in good agreement with the in-flight measurements of the ambient fields inside storm-cloud (Lalande et al., 1999).

Furthermore, the simulations demonstrated that the electric field inside the radome is not reduced significantly by the radome wall itself, which indicates that the radome does not produce screening effects. This shows that leader can start from the internal electrode (radar antenna) causing flashover across the radome. Therefore, strips to produce the screening effect must be used to avoid the initiation of streamers from the antenna. The lightning strike to the radome does not damage the radome surface if discharges do not occur from the metal parts inside the radome. This points out that the main purpose of the protection system should be the screening (shielding) of the electric field inside the radome. Poor shielding permits the inception of a discharge from the internal electrode, so the solid strips are usually used to produce the shielding effect. However, effective shielding has undesirable interference effects on antenna radiation. Therefore, the optimal length and number of the strips should be determined.

Significant shielding effect is created by water and ice layers on the radome surface. Under these conditions, the lightning leader can be initiated from the radome tip. Note that the dielectric constant values of ice depend on the frequency of the external field or the rate of voltage rise, and these values affect the electric field magnitude. For example, the values of $\varepsilon_{ice} = 5$ for $1000\ kV/\mu s$ and $\varepsilon_{ice} = 70$ for $10\ kV/\mu s$ were used in (Hardwick et al., 2003). This work has shown that the ice layer does not screen the high frequency radiation associated with the radar.

In high ambient humidity conditions (>60%), the radome becomes moderately conductive because of humidity absorption at its surface (Ulmann et al., 2001; Delannoy et al., 2001). Although this decreases the internal field due to shielding effect, it also reduces the efficiency of the strips.

Numerical simulations have shown that the shielding effect is produced only by solid strips, there is no practical shielding by segmented strips in the absence of a discharge. It was demonstrated that the field intensification area is forced out from the metal electrode (antenna) surface to the front of the radome, thereby preventing discharge initiation from the antenna. However, too strong shielding of antenna surface by increasing the number and the length of strips can cause the field intensification at the frontal area of the radome which can be sufficient to initiate the discharge. Hence, the shielding of the antenna surface as much as possible is not the best solution to the problem. It is necessary to optimize the electric field distribution with respect to the streamer and leader discharge initiation conditions.

Both the fast and slow waveforms (MIL STD 1757 Waveforms A and D respectively) are used for testing radomes (Ulmann et al., 1999). Waveform A has $1000\ kV/\mu s$ rate of rise, and Waveform D has 50-250 μs rise time. It was concluded (Ulmann et al., 1999) that Waveform D represents the in-flight environment more accurately than Waveform A. For aircraft intercepting approaching leaders, rates of rise of the electric field, dU/dt of 10^8 to 10^{10} $V/m/s$ were estimated (Lalande et al., 1999) at the aircraft. If 1 $MV/\mu s$ (waveform A) is applied over a 1m gap, this will give $dU/dt \approx 10^{12}$ $V/m/s$. Hence, the slower voltage Waveform D tests might be more appropriate. In our tests, we have $dU/dt \approx U/\tau_f/L \approx 2.8 \cdot 10^5 V/2 \cdot 10^{-6} s/0.7m \approx 2 \cdot 10^{11}$ $V/m/s$. However, the voltage rise time is important when the voltage is applied directly to the strip. If the high-voltage electrode is placed far from the strip, the breakdown process of the strip is determined by the field generated by the ionization front of the discharge, i.e. by the space charge of the streamers. The magnitude of this field is affected by the velocity of the streamer/leader ionization front, but not by the applied voltage waveform.

Besides direct strikes to aircraft radome, the aircraft could be subjected to indirect strikes. Lightning strike entrance and exit points are usually found at sharp structures of the aircraft, around which the electric field enhancement takes place, but also can occur at any part of the aircraft, including the fuselage, stabilisers, antennas, etc. Observations of such strikes were conducted in a laboratory experiments with aircraft models (Chernov et al., 1992; Petrov et al., 1996). It is seen from Fig. 19, that the nose radome can also be exit point of lightning strike depending on the aircraft position with respect to the approaching lightning threat.

It is worth highlighting here that the lightning diverter strips concept could be adapted for use in protection of ground antennas for ultra-high-frequency communications, which are difficult to protect from direct lightning strikes because interference to the radiation field arises when standard air-terminal shielding is installed (Bruel et al., 2004).

Fig. 19. Laboratory testing of lightning strikes to an aircraft model.

7. Conclusion

Theoretical analysis and numerical simulations together with experimental laboratory tests of lightning discharge interaction with aircraft radome demonstrated the applicability of existing lightning attachment models to create optimal protection systems against lightning strikes.

The following points can be concluded from the analysis:

i.   Electric field intensification by aircraft flying at high altitudes exceeds the threshold to initiate the lightning leader (50-100 kV/m), this explains why about 90% of lightning strikes to aircraft are initiated by the aircraft.

ii. The shielding effect of dielectric radome material itself is less than 10%, so the lightning leader can be initiated from the radar antenna.

iii. The penetration of the electric field, created by the lightning channel or storm-cloud, into the radome is significantly decreased by ice and/or water layers on the radome surface; however, this may cause also the occurrence of punctures.

iv. Strong diversion effect for the strips comprising isolated metal disks or rings is observed for positive as well as for negative polarity discharges; this type of diverter strip can be used together with the solid strips in order to decrease the interference effect on antenna radiation.

v.   Numerical simulations have shown strong radar shielding effects produced by solid strips and no practical shielding by segmented strips in the absence of a discharge.

8. Acknowledgment

N.I.P. and G.N.P. thank colleagues of the High Voltage Group of the Cardiff School of Engineering for hospitality while they worked as guests in their laboratory.

9. References

Amason, M.; et al. (1975). Aircraft application of segmented-strip lightning protection systems, *Proceedings of Conf. on Lightning and Static Electricity*, pp.1-14, London, UK, 1975

Bruel, C.; Barilleau, D. & Rousseau, A. (2004). Application of aircraft lightning protection to radar stations, *Proceedings of 27th Int. Conf. on Lightning Protection*, pp. 975-977, Avignon, France, September 13-16, 2004

Chernov, E.; Lupeiko, A. & Petrov, N. (1992). Repulsion effect in orientation of Lightning discharge. *J.de Phys. III*, Vol. 2, (July 1992), pp. 1359-1365

Clifford, D. & Casemir, H. (1982). Triggered lightning. *IEEE Trans. Electromagnetic Compatibility*, Vol. 21, (January 1982), pp. 112-122, ISSN 0018-9375

Cooray, V.; Rakov, V. & Theethayi, N. (2004). The relationship between the leader charge and the return stroke current – Berger's data revisited, *Proceedings of 27th Int. Conf. on Lightning Protection*, pp. 145-150, Avignon, France, September 13-16, 2004

Delannoy, A.; Bondiou-Clergerie, A.; Lalande, P.; et.al. (2001). New investigations of the mechanisms of lightning strike to radomes Part II: Modeling of the protection efficiency, *Proceedings of Int. Conf. on Lightning and Static Electricity*, paper No 2001-01-2884, Seattle, USA, September 11-13, 2001

Handbook of Chemistry and Physics. (2001). CRC Press, ISBN 0849304822

Hardwick, J.; Plumer, A. & Ulmann, A. (1999). Review of the joint radome programme, *Proceedings of ICOLSE'99*, pp.59-65, ISBN 0768003938, Toulouse, France, June 22-24, 1999

Hardwick, C.; Hawkins, K. & Sanders, M. (2003). Effect of water and icing on segmented diverter strip performance, *Proceedings of ICOLSE'03*, pp. 80.1-80.8, ISBN 1857681525, 9781857681529, Blackpool, UK, September 16-18, 2003

Larsson, A. (2002). The interaction between a lightning flash and an aircraft in flight. *C.R. Physique*, Vol 3, (December 2002), pp. 1423-1444

Lalande, P.; Bondiou-Clergerie, A. & Laroche, P. (1999). Analysis of available in-flight measurements of lightning strikes to aircraft, *Proceedings of ICOLSE'99*, pp.401-408, Toulouse, France, June 22-24, 1999

Mazur, V. (1989). Triggered lightning strikes to aircraft and natural intracloud discharges. *J. Geophys. Res.*, Vol. 94, (March 1989), pp. 3311-3325, ISSN 0148-0227

Mazur, V. (1989). A physical model of lightning initiation on aircraft in thunderstorms. *J. Geophys. Res.*, Vol. 94, (March 1989), pp. 3326-3340, ISSN 0148-0227

Mazur, V.; Fisher, B. & Brown, P. (1990). Multistroke cloud-to-ground strike to the NASA F-106B airplane, *J. Geophysical Research*, Vol. 95, no. D5, (May 1990), pp. 5471-5484, ISSN 0148-0227

Mazur, V. & Moreau, J. (1992). Aircraft-triggered lightning: processes following strike initiation that affect aircraft, *J. Aircr.*, Vol. 29, (August 1992), pp. 575-580, ISSN 0021-8669

Marshall, T. & Rust, W. (1991). Electric field soundings through thunderstorms. *J. Geophys. Res.*, Vol. 96(22), (December 1991), pp. 297-306, ISSN 0148-0227

Petrov, N. & Waters, R. (1994). Conductor height and altitude: effect on striking distance, *Proc. Int. Conf. Lightning and Mountains*, pp. 52-57, SEE, Chamonix-Mont-Blanc, June 6-9, 1994.

Petrov, N. & Waters, R. (1995). Determination of the striking distance of lightning to earthed structures, *Proc. R. Soc. Lond. A*, Vol. 450, No. 1940, (September 1995), pp. 589-601, ISSN 1471-2946

Petrov, N.; Avansky, V.; Efimova, N. & Petrova, G. (1996). Experimental and theoretical investigations of the orientation of leader discharge to isolated and earthed objects, Proceedings of 23th Int. Conf. on Lightning Protection, pp.254-259, Vol.1, Firenze, Italy, September 23-27, 1996

Petrov, N. & D'Alessandro, F. (2002). Theoretical analysis of the processes involved in lightning attachment to earthed structures, *J. Phys. D: Appl. Phys.*, Vol. 35, No. 14, (July 2002), pp. 1788-1795, ISSN 0022-3727

Petrov, N. & Waters, R. (1999). Striking distance of Lightning to earthed structures: effect of stroke polarity, *Proc. 11th Int. Symp. on High Voltage Engineering*, pp.220-223, Vol. 2, London, UK, August 23-27, 1999

Plumer, J. & Hoots, L. (1978). Lightning protection with segmented diverters, *Proceedings of IEEE Int. Symp. Electromagnetic Compatibility*, pp.196-203, 1978

Ulmann, A.; Hardwick, J. & Plumer, A. (1999). Laboratory Reproduction of In-Flight Failures of Radomes, *Proceedings of ICOLSE'99*, pp.493-496, ISBN 0768003938, Toulouse,France, June 22-24, 1999

Ulmann, A.; Brechet, P.; Bondiou-Clergerie, A.; et.al. (2001). New investigations of the mechanisms of lightning strike to radomes Part I: Experimental study in high voltage laboratory, *Proceedings of Int. Conf. on Lightning and Static Electricity*, paper No 2001-01-2883, Seattle, USA, September 11-13, 2001

Uman, M. & Rakov, V. (2003). The interaction of lightning with airborne vehicles. *Progress in Aerospace Sciences*, Vol.39, No 1, (January 2003), pp. 61-81, ISSN 0376-0421

10

Synthetic Aperture Radar Systems for Small Aircrafts: Data Processing Approaches

Oleksandr O. Bezvesilniy and Dmytro M. Vavriv
Institute of Radio Astronomy of the National Academy of Sciences of Ukraine
Ukraine

1. Introduction

The synthetic aperture radar (SAR) is considered now as the most effective instrument for producing radar images of ground scenes with a high spatial resolution. The usage of small aircrafts as the platform for the deployment of SAR systems is attractive from the point of view of many practical applications. Firstly, this enables for a substantial lowering of the exploitation costs of SAR sensors. Secondly, such solution provides a possibility to perform a rather quick surveillance and imaging of particular ground areas. Finally, the progress in this direction will allow for a much wider application of SAR sensors.

However, the formation of high-quality SAR images with SAR systems deployed on small aircrafts is still a challenging problem. The main difficulties come from significant variations of the aircraft trajectory and the antenna orientation during real flights. These motion errors lead to defocusing, geometric distortions, and radiometric errors in SAR images.

In this chapter, we describe three effective approaches to the SAR data processing, which enable the solution of the above problems:

1. Time-domain SAR processing with clutter-lock and geometric correction by resampling,
2. Time-domain SAR processing with built-in geometric correction and multi-look radiometric correction,
3. Range-Doppler algorithm with the 1-st and 2-nd order motion compensation.

The proposed solutions have been successfully implemented in Ku- and X-band SAR systems developed and produced at the Institute of Radio Astronomy of the National Academy of Sciences of Ukraine. The efficiency of the proposed algorithms is illustrated by SAR images obtained with these SAR systems.

The chapter is organized as follows. In Section 2, basic principles of SAR data processing is described. In Section 3, the problem of motion errors of airborne SAR systems is considered, and the appearance of geometric distortions and radiometric errors in SAR images is discussed. The three data processing approaches are considered in details in Sections 4, 5, and 6. Section 7 describes the RIAN-SAR-Ku and RIAN-SAR-X systems used in our experiments. The conclusion is given in Section 8.

2. Principles of SAR data processing

The synthetic aperture technique is used to obtain high-resolution images of ground surfaces by using a radar with a small antenna installed on an aircraft or a satellite. The radar pulses backscattered from a ground surface and received by the moving antenna can be considered as the pulses received by a set of antennas distributed along the flight trajectory. By coherent processing of these pulses it is possible to build a long virtual antenna – the synthetic aperture that provides a high cross-range resolution. A high range resolution is typically achieved by means of a pulse compression technique that involves transmitting long pulses with a linear frequency modulation or a phase codding.

2.1 Concept of the synthetic aperture

Practical SAR systems are produced to operate in one or several operating modes. Depending on the mode, they are referred as the strip-map SAR, the spot-light SAR, the inverse SAR, the ScanSAR, and the interferometric SAR (Bamler & Hartl, 1998; Carrara at al., 1995; Cumming & Wong, 2005; Franceschetti & Lanari, 1999; Rosen at al., 2000; Wehner, 1995). We shall consider mainly the most popular and practically useful strip-map SAR operating mode. However, the presented further results are applicable to other modes to a large extent.

In the strip-map SAR mode, the radar performs imaging of a strip on the ground aside of the flight trajectory. Geometry of the strip-map mode is shown in Fig. 1. The aircraft flies along the straight line above the x-axis with the velocity V at the altitude H above the ground plane (xy).

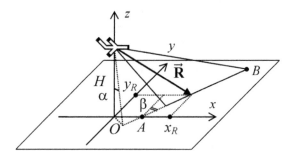

Fig. 1. Geometry of the strip-map SAR mode.

The orientation of the real antenna beam is described by the pitch angle α and the yaw angle β that are measured with respect to the flight direction. The line AB in Fig. 1 is the intersection of the elevation plane of the real antenna pattern and the ground plane. This line is called the Doppler centroid line. The coordinates of the point (x_R, y_R) on this line at the slant range R from the aircraft are given by

$$x_R = H \tan\alpha \cos\beta + \sin\beta \sqrt{R^2 - H^2 - (H\tan\alpha)^2} ,$$ (1)

$$y_R = -H \tan \alpha \sin \beta + \cos \beta \sqrt{R^2 - H^2 - (H \tan \alpha)^2} . \tag{2}$$

In order to form the synthetic aperture and direct the synthetic beam to the point (x_R, y_R), the signal $s_R(\tau + t)$ backscattered from this point should be summed up coherently on the interval of synthesis $-T_S / 2 \leq \tau \leq T_S / 2$ taking into account the propagation phase $\varphi(\tau)$ (Cumming & Wong, 2005; Franceschetti & Lanari, 1999):

$$I(t, x_R, y_R) = \left| \frac{1}{T_S} \int_{-T_S/2}^{T_S/2} s_R(\tau + t) h_R(\tau) d\tau \right|^2 , \tag{3}$$

$$h_R(\tau) = w_R(\tau) \exp[-i\varphi(\tau)], \quad \varphi(\tau) = -\frac{4\pi}{\lambda} R(\tau) . \tag{4}$$

Here $I(t, x_R, y_R)$ is the SAR image pixel, t is the time when the aircraft is at the centre of the synthetic aperture $(0, 0, H)$, τ is the time within the interval of synthesis, $h_R(\tau)$ is the azimuth reference function in the time domain, $w_R(\tau)$ is the weighting window applied to improve the side-lobe level of the synthetic aperture pattern, λ is the radar wavelength, and $R(\tau)$ is the slant range to the point:

$$R(\tau) = \sqrt{(x_R - V\tau)^2 + y_R^2 + H^2} = \sqrt{R^2 - 2x_R V\tau + (V\tau)^2} . \tag{5}$$

If the slant range $R(\tau)$ changes during the time of synthesis T_S more than the size of the range resolution cell, then the target signal "migrates" through several range cells. This effect known as the range migration should be taken into account during the aperture synthesis. The one-dimensional backscattered signal $s_R(\tau + t)$ should be obtained from the two-dimensional "azimuth – slant range" matrix of the range-compressed radar data by the interpolation along the migration curve (5).

The instant Doppler frequency of the received signal is, approximately,

$$f(\tau) = -\frac{2}{\lambda} \frac{dR(\tau)}{dt} \approx F_{DC} + F_{DR}\tau , \tag{6}$$

where the Doppler centroid F_{DC} and the Doppler rate F_{DR} are given by

$$F_{DC} = \frac{2}{\lambda} V \frac{x_R}{R} , \tag{7}$$

$$F_{DR} = -\frac{2}{\lambda} \frac{V^2}{R} \left[1 - \left(\frac{x_R}{R} \right)^2 \right] . \tag{8}$$

It is useful to note that the Doppler centroid determines the synthetic beam direction, whereas the Doppler rate is responsible for the beam focusing.

From the point of view of signal processing, the formation of the synthetic aperture (3) is the matched filtering of linear frequency modulated signals (6). Such filtering can be performed either in the time or in the frequency domain. Accordingly, there are time- and frequency-domain SAR processing algorithms.

It is easy to show that the azimuth resolution ρ_X is given by (Cumming & Wong, 2005; Carrara at al., 1995)

$$\rho_X = K_w \frac{V}{\Delta F_D} \, . \tag{9}$$

Here $\Delta F_D = |F_{DR}| T_S$ is the Doppler frequency bandwidth that corresponds to the interval of the synthesis. The coefficient K_w describes the broadening of the main lobe of the synthetic aperture pattern caused by windowing.

In order to improve the quality of SAR images, a multi-look processing technique is used in most modern SAR systems (Moreira, 1991; Oliver & Quegan, 1998). According to such technique, a long synthetic aperture is divided on shorter intervals that are processed independently to build several SAR images of the same ground scene, called SAR looks. It can be considered as building the synthetic aperture with multiple synthetic beams. A non-coherent averaging of the SAR looks into one multi-look image is used to reduce speckle noise and to reveal fine details in SAR images. Multi-look processing can be used for other applications, for example, for measuring the Doppler centroid with a high accuracy and high spatial resolution and retrieving 3D topography of ground surfaces (Bezvesilniy et al., 2006; Bezvesilniy et al., 2007; Bezvesilniy et al., 2008; Vavriv & Bezvesilniy, 2011b).

In the next sections we consider peculiarities of the realization of SAR processing algorithms in time and frequency domains.

2.2 SAR processing in time domain

The SAR processing in the time domain is performed according to the relations (3)-(5). The block-scheme of the algorithm is shown in Fig. 2.

The received range-compressed radar data are stored in a memory buffer. The buffer size in the range corresponds to the swath width; the buffer size in the azimuth is determined by the time of synthesis. The basic step of the SAR processing procedure for a given range R includes the following calculations:

1. Calculation of the Doppler centroid (7), the Doppler rate (8), and the required time of synthesis (9),
2. Interpolation along the migration curve (5),
3. Multiplication by the reference function with windowing (4), and
4. Coherent summation (3).

As the result, a single pixel of the SAR image is obtained representing the ground point on the Doppler centroid line at the range R. This basic step is repeated for all ranges within the swath producing a single line of the SAR image in the range direction. In order to form the next line of the SAR image, the data in the buffer is shifted in the azimuth and supplemented with new data, and the computations are repeated.

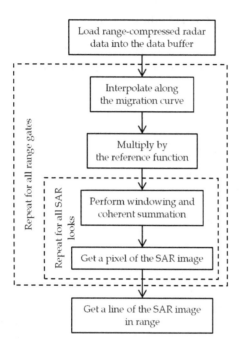

Fig. 2. SAR processing in time domain.

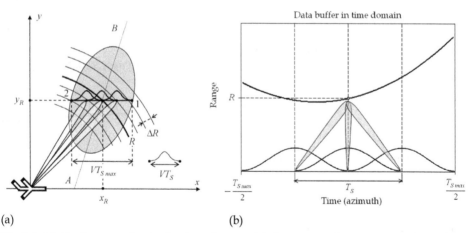

(a)   (b)

Fig. 3. Multi-look processing in the time domain: (a) the antenna footprint consideration, (b) the data buffer consideration.

The multi-look processing in the time domain is usually performed directly following the definition (Moreira, 1991; Oliver & Quegan, 1998). Namely, the reference functions and range migration curves are built for the long interval of synthesis $T_{S\max}$, which is the time required for the ground target to cross the antenna footprint from point 1 to point 2 in Fig.

3a. The multi-look processing is performed by splitting the long interval of the synthesis T_{Smax} on several sub-intervals T_S, forming in this manner multiple synthetic beams pointed to the same point on the ground at the different moments of time, as shown in Fig. 3b. The number of the looks for a scheme with the half-overlapped sub-intervals is given by

$$N_L = \text{int}\left\{\frac{T_{Smax}}{T_S/2}\right\} - 1. \tag{10}$$

2.3 SAR processing in frequency domain

The SAR data processing can be also performed effectively in the frequency domain. It is known that the convolution of two signals in the time domain is equivalent to the multiplication of their Fourier pairs in the frequency domain. The corresponding computations are efficient due to the application of the fast Fourier transform (FFT). A number of FFT-based SAR processing algorithms have been so far developed (Cumming & Wong, 2005).

In particular, the range-Doppler algorithm (RDA) (Cumming & Wong, 2005) is a relatively simple and widely-used FFT-based algorithm. The processing steps of this algorithm are shown in Fig. 4 and illustrated also in Fig. 5. The received radar data are stored in a large memory buffer. The buffer size in the range direction corresponds to the swath width, and the buffer size in the azimuth direction is equal to the length of the FFT that covers many intervals of synthesis. First, the range-compressed data are transformed into the range-Doppler domain by applying FFT in the azimuth. The frequency scale is limited by the pulse repetition frequency (PRF). Then, the range migration correction is performed in the frequency domain. By using the relation (6) between the instant frequency f and the time τ within the interval of synthesis (preserving the square-root law for the slant range) one can derive the formula for the migration curve in the frequency domain from the migration curve (5) in the time domain:

$$R(f) = R\sqrt{1 - \frac{\lambda^2 F_{DC}^2}{4V^2}} \Big/ \sqrt{1 - \frac{\lambda^2 f^2}{4V^2}}. \tag{11}$$

After that, the phase compensation and windowing are applied for the azimuth compression. By using the principle of stationary phase (Cumming & Wong, 2005) an expression for the reference function in the frequency domain is obtained:

$$h_R(f) = w_R(f)\exp[-i\theta(f)], \ \theta(f) = -\frac{4\pi}{\lambda}R\left[\sqrt{1 - \left(\frac{\lambda f}{2V}\right)^2}\sqrt{1 - \left(\frac{\lambda F_{DC}}{2V}\right)^2} + \left(\frac{\lambda f}{2V}\right)\left(\frac{\lambda F_{DC}}{2V}\right)\right]. \tag{12}$$

Finally, the SAR image is formed by applying the inverse FFT in the azimuth. Thus, the basic processing step in the frequency domain performed for a given range gives the line of the SAR image in the azimuth. This basic step is repeated for all ranges within the swath producing the complete SAR image of the ground scene presented in the data frame.

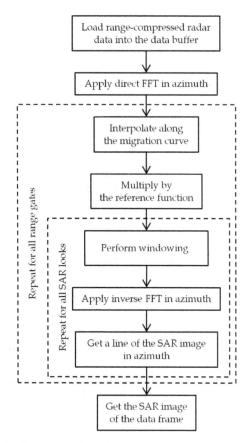

Fig. 4. SAR processing in frequency domain.

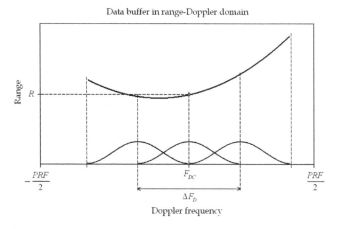

Fig. 5. Multi-look processing in frequency domain.

In the cases of a significantly squinted geometry (a large antenna yaw angle) or a very high resolution, or a large number of looks, an additional processing step called "secondary range compression" is required (Cumming & Wong, 2005).

The multi-look processing is performed in the frequency domain by dividing the whole Doppler band $\Delta F_{D\max} = |F_{DR}| T_{S\max}$ of the backscattered radar signals into the sub-bands ΔF_D for the separate azimuth compression (Cumming & Wong, 2005; Carrara at al., 1995):

$$\Delta F_D = \frac{\Delta F_{D\max}}{(N_L + 1)/2}. \tag{13}$$

For the multi-look processing scheme with the half-overlapped sub-bands, the central frequencies of the SAR looks with respect to the Doppler centroid are given by

$$\Delta F_C(R, n_L) = F_{DC}(R) - n_L \frac{\Delta F_D}{2}, \tag{14}$$

where $n_L = -N_L / 2, ..., N_L / 2 - 1$ is the SAR look index. Since the Doppler rate (8) is always negative, the first sub-interval in the time domain corresponds to the last sub-band in the frequency domain. Therefore, we write the minus sign in (14).

3. Problem of aircraft motion errors

Deviations of the aircraft flight trajectory and instabilities of the aircraft orientation significantly complicate the formation of SAR images. Such motion errors lead to defocusing, geometric distortions, and radiometric errors in SAR images (Blacknell et al., 1989; Buckreuss, 1991; Franceschetti & Lanari, 1999; Oliver & Quegan, 1998). In this section, we shall discuss these problems and their solutions in details.

3.1 Aircraft flight with motion errors

The trajectory of an aircraft may deviate from a straight line significantly in real flights. The orientation of the aircraft could also be unstable. These motion errors should be measured and compensated in order to produce high-quality SAR images. We assume that the navigation system is capable of measuring the aircraft trajectory and the aircraft velocity vector. We suppose also that the orientation of the real antenna beam with respect to the velocity vector is known.

Usually, the final product of the strip-map SAR system is a sequence of SAR images of a particular dimension, built in a projection to the ground plane, with indication of the north direction and the latitude-longitude position. Later, if necessary, several consequent images can be stitched together to produce a larger map of a particular ground area of interest. Thus, the received radar data is processed by data frames. Each frame gives one SAR image from the image sequence. The data frames are usually overlapped to guarantee successful stitching of the produced SAR images without gaps.

In order to produce the SAR image from the data frame, it is needed to define a reference flight line for this frame, the averaged flight altitude H_{ref} and the averaged velocity V_{ref}. Under unstable flight conditions, the reference flight line should be close to the actual curvilinear flight trajectory of the aircraft. Also, the reference antenna pitch and yaw angles α_{ref} and β_{ref}, which describes the averaged orientation of the real antenna beam during the time of the data frame acquisition, should be introduced.

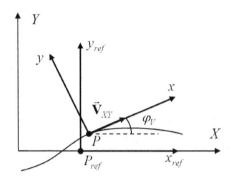

Fig. 6. The scene coordinate system, the reference local coordinate system, and the actual local coordinate system.

Let us define the scene coordinate system (X, Y, Z) so that the reference flight line goes exactly above the X axis. The final-product SAR image is to be sampled on the coordinate grid of the ground plane (X, Y) of this coordinate system. The scene coordinate system is shown in Fig. 6 together with the actual local coordinate system (x, y, z), which slides along the real aircraft flight trajectory, and the reference local coordinate system $(x_{ref}, y_{ref}, z_{ref})$, which slides along the X axis (that is along the reference flight line). The current flight direction is described by the angle φ_V between the horizontal component of the velocity vector $\vec{V}_{XY}$ and the X axis.

The aircraft trajectory $(X_A(t), Y_A(t), Z_A(t))$ is described in the scene coordinate system. The actual local coordinates (x, y) and the reference local coordinates (x_{ref}, y_{ref}) are related to each other as follows:

$$x = [x_{ref} - X_A(t) + V_{ref}t]\cos\varphi_V(t) + [y_{ref} - Y_A(t)]\sin\varphi_V(t), \tag{15}$$

$$y = -[x_{ref} - X_A(t) + V_{ref}t]\sin\varphi_V(t) + [y_{ref} - Y_A(t)]\cos\varphi_V(t). \tag{16}$$

The pitch $\alpha(t)$ and yaw $\beta(t)$ angles describe the antenna beam orientation with respect to the current aircraft velocity vector or, in other words, with respect to the actual local coordinate system. It means that when the synthetic beam is directed to the point (x_R, y_R) on the Doppler centroid line by using the Doppler centroid (7), the Doppler rate (8), and the migration curve (5) under unstable flight conditions, the coordinates (x_R, y_R) are given in the actual local coordinate system. In order to find the scene coordinates (or the reference local coordinates) of this point, the above relations (15), (16) should be used.

An example of motion errors typical for a light-weight aircraft AN-2 is shown in Fig. 7. In the figure, one can see the coordinate grid of the radar coordinates "slant range – azimuth" projected onto the ground plane (X, Y). The horizontal curves are the curves of the constant slant range from the aircraft. They are curved because of deviations of the trajectory from the straight line. The vertical lines are the central lines of the antenna footprint (the Doppler centroid lines) for the consequent aircraft positions. As it is seen, the central lines are not equidistant and not parallel because of variations of the antenna orientation.

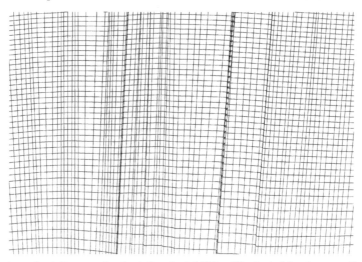

Fig. 7. Trajectory deviations and orientation instabilities illustrated by the coordinate grid in the radar coordinates "slant range – azimuth" on the ground plane.

3.2 Geometric distortions in SAR images

The direction of the synthetic beam is determined by the used Doppler centroid with respect to the current velocity vector. In other words, the Doppler centroid controls the direction of the synthetic beam with respect to the actual local coordinate system. Therefore, if the deflections of the velocity vector from the reference flight direction (described by the angle φ_V in Fig. 6) are not compensated properly, then the synthetic beam is moving forward or backward along the flight path with respect to the scene coordinate system. It means that the scene will be sampled non-uniformly in the azimuth direction resulting in geometric distortions in SAR images. For example, if the synthetic beams are pointed to the centre of the real beam, i.e. to the Doppler centroid line, then the scene will be sampled on a non-uniform grid like that shown in Fig. 7.

If the aircraft trajectory and the orientation of the synthetic aperture beams are known, the geometric distortions can be corrected by resampling of the obtained SAR images to a rectangular grid on the ground plane. This resampling procedure is described in Section 4. However, this approach could be inefficient in the case of significant geometric distortions.

Alternatively, geometric errors can be avoided if the orientation of the synthetic beams is adjusted at the stage of synthesis by using the trajectory information. The purpose of this adjustment is to keep the beam orientation constant with respect to the reference flight direction. This is the idea of the built-in geometric correction discussed in Section 5.

The correction of the phase errors and range migration errors caused by trajectory deviations can be applied to the raw data before the aperture synthesis. After such compensation, the raw data look like be collected from the reference straight line. After such motion compensation, the synthetic beams will be set with respect to the reference local coordinate system. Such approach is widely used with the SAR processing algorithms working in the frequency domain. This motion compensation technique is considered in Section 6 with application to the range-Doppler algorithm.

3.3 Radiometric errors in SAR images

The problem of radiometric errors is illustrated in Fig. 8. If there are no orientation errors, the synthetic beam of the central look is directed to the centre of the real antenna beam, and all SAR look beams are within the main lobe of the real antenna pattern, as shown in Fig. 8a. The antenna orientation errors lead to the situation when the SAR beams are directed outside the real antenna beam to not-illuminated ground areas, as shown in Fig. 8b, resulting in radiometric errors.

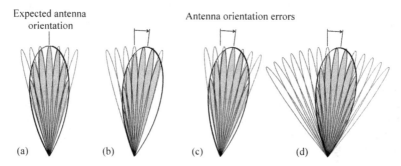

Fig. 8. Multi-look processing without antenna orientation errors (a) and with orientation errors: (b) without clutter-lock, (c) with clutter-lock, (d) with extended number of looks.

Instabilities of the aircraft orientation can be compensated by the antenna stabilization by mounting it on a gimbal. It helps to keep the constant antenna beam orientation. However, this approach is rather complicated and expensive.

The application of a wide-beam antenna firmly mounted on the aircraft is a less expensive way to guarantee a uniform illumination of the ground scene despite of instabilities of the platform orientation. Several shortcomings of this approach should be admitted. The application of a wide antenna beam means some degradation of the radar sensitivity. Also, it calls for a higher PRF to sample the increased Doppler frequency band. Moreover, only the central part of the antenna footprint will be illuminated uniformly limiting the number of looks that can be built without an additional radiometric compensation.

The clutter-lock technique (Li at al., 1985; Madsen, 1989) is usually used to avoid radiometric errors in SAR images. According to the clutter lock technique, the azimuth reference functions are built adaptively so that the synthetic beams track the direction of the real antenna beam staying within the main lobe of the real antenna pattern as shown in Fig. 8c. However, the variations of the synthetic beam orientation due to the clutter-lock naturally lead to geometric distortions in SAR images.

The clutter lock technique is effective if the variations of the antenna beam orientation are slow in time and small as compared to the real antenna beam width in the azimuth. In this case, the geometric distortions can be corrected by re-sampling. Provided that the orientation instabilities are fast and significant, the clutter-lock leads to strong geometric distortions in SAR images which cannot be easily corrected by re-sampling.

We have proposed an alternative radiometric correction approach, which is based on a multi-look SAR processing with an extended number of SAR looks (Bezvesilniy et al., 2010c; Bezvesilniy et al., 2010d; Bezvesilniy et al., 2011b; Bezvesilniy et al., 2011c). This technique can be used instead of the clutter-lock. The idea of the approach consists in the formation of an extended number of looks to cover directions beyond the main lobe of the real antenna pattern as illustrated in Fig. 8d. In such approach, some of the SAR look beams are always presented within the real antenna beam despite of the orientation errors. In Section 5, we describe how to combine these extended SAR looks to produce the multi-look SAR image without radiometric errors. This approach is appropriate for the cases when the clutter-lock cannot be applied because of fast orientation instabilities or for SAR processing algorithms that cannot be used together with the clutter-lock. The proposed method also allows correcting the radiometric errors in SAR images if the antenna orientation is not known accurately.

3.4 Dilemma: geometric distortions vs. radiometric errors

From the above considerations, one can conclude that an attempt to avoid geometric errors by the appropriate pointing of the synthetic beams leads to radiometric errors. And vice versa, the clutter-lock results in geometric errors. So the dilemma of "geometric distortions vs. radiometric errors" should be resolved when developing any SAR data processing approach for SAR systems with motion errors.

We describe three alternative approaches to this problem. In the first approach, described in Section 4, the priority is set to avoiding radiometric errors and the clutter-lock is applied. Geometric errors are corrected by resampling of the obtained SAR images. In the second approach, considered in Section 5, the geometric accuracy of SAR images is the primary goal and we implement a synthetic beam control algorithm called "built-in geometric correction" to point the beams to the nodes of a correct rectangular grid on the ground plane. Radiometric errors are corrected by multi-look processing with extended number of looks. In the third approach, discussed in Section 6, a range-Doppler algorithm with the 1-st and 2-nd order motion compensation is considered, which allows obtaining SAR images without significant geometric errors. The application of a wide-beam real antenna could be a solution of the problem of radiometric errors for this approach.

4. Time-domain SAR processing with clutter-lock and geometric correction by resampling

In this section, we consider a time-domain SAR data processing algorithm assuming that the aircraft flight altitude and velocity, as well as the antenna beam orientation angles are changed slowly in the sense that they can be considered constant during the time of the synthesis. The main steps of the algorithm are the same as in the case of the straight-line motion with a constant orientation. These steps are described in the block-scheme shown in

Fig. 2. At each step of the synthesis, the reference function and migration curves are adjusted according to the estimated orientation angles of the real antenna beam providing the clutter-lock. Due to the clutter-lock, it is possible to avoid radiometric errors. Geometric errors are corrected by resampling of the obtained SAR images on the post-processing stage.

4.1 Estimation of the antenna orientation angles from Doppler centroid measurements

According to the clutter-lock technique, the synthetic beams are built adaptively to track the direction of the real antenna beam. The orientation angles of the aircraft can be measured by a navigation system. The commonly used navigation systems are based on Inertial Measurement Unit (IMU) or on a combination of IMU and attitude GPS. They are typically rather expensive and do not always provide the required accuracy and the needed rate of measurements. We have proposed an effective method for the estimation of the antenna orientation angles – pitch and yaw – from the Doppler measurements. The application of this technique has allowed us to simplify the navigation system by reducing it to a simple GPS receiver to measure the platform velocity and coordinates only.

The mathematical background of this technique is as follows. The dependence of the Doppler centroid on the slant range is given by

$$F_{DC} = \frac{2}{\lambda} \frac{(\mathbf{R} \cdot \mathbf{V})}{R} = \frac{2}{\lambda} \frac{x_R V_x - H V_z}{R} . \tag{17}$$

The slant range vector $\mathbf{R} = (x_R, y_R, -H)$ is directed from the antenna phase centre to the point (x_R, y_R) on the Doppler centroid line as shown in Fig. 1, and the aircraft velocity vector is $\mathbf{V} = (V_x, 0, V_z)$. In opposite to (7), formula (17) accounts for the possible vertical direction of the aircraft motion. Substituting in (17) the expression (1) for the coordinate of a point on the Doppler centroid line, we rewrite the above dependence as

$$F_{DC}(R, \alpha, \beta) = \frac{2}{\lambda} \frac{V_x}{R} \left[H \tan \alpha \cos \beta + \sin \beta \sqrt{R^2 - H^2 - (H \tan \alpha)^2} - H \frac{V_z}{V_x} \right]. \tag{18}$$

The behaviour of the Doppler centroid on range depends strongly on particular values of the antenna pitch and yaw angles as illustrated in Fig. 9. It means that theoretically the antenna beam orientation angles can be estimated via an analysis of the dependence of the measured Doppler centroid on range. However, for the practical implementation of this idea, it was needed to answer the questions: Would it be a reliable estimate? Is it possible to achieve the required accuracy of the angle measurements? And, is it possible to realize this estimation in real time? Fortunately, we have found solutions which provide positive answers on the above questions.

We have found that the pitch and yaw angles can be estimated by fitting the theoretical dependence of the Doppler centroid on range (18) into a set of Doppler centroid values $F_{DC}^{[n]} = F_{DC}(R_n)$ roughly estimated from the received data at each range gate from the Doppler spectra calculated by using the FFT. Here n is the range gate index.

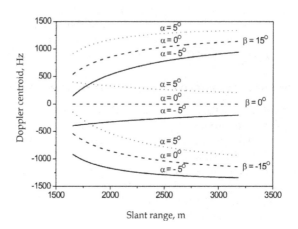

Fig. 9. Dependence of the Doppler centroid on slant range.

We have developed the following fast and effective fitting procedure. By introducing new variables X_n^i, Y_n as

$$X_n^i = \sqrt{R_n^2 - H^2 - (H \tan \alpha_{i-1})^2} \ , \ Y_n = \frac{\lambda F_{DC}^{[n]}}{2V_x} R_n + H \frac{V_z}{V_x} \ , \tag{19}$$

the dependence (18) is transformed into the equation of a straight line:

$$Y_n = (H \tan \alpha_i \cos \beta_i) + X_n^i (\sin \beta_i) \ . \tag{20}$$

Thus, the problem of fitting of the non-linear dependence (18) is turned into the well-known task of fitting of a line into a set of experimental points. The only difficulty is that the unknown pitch angle appears in the transformation of the coordinates (19). We have solved this difficulty by using an iteration procedure. The fitting is performed iteratively with respect to the pitch angle considered as a small parameter. The index $i = 1, 2, 3, \dots$ in (19), (20) is the iteration index. At the first iteration, the pitch angle is assumed to be zero: $\alpha_0 = 0$.

It has been found that two iteration are typically enough to achieve the required accuracy of about 0.1° in real time. The method has been implemented in SAR systems developed and produced at the Institute of Radio Astronomy (Vavriv at al., 2006; Vavriv & Bezvesilniy, 2011a; Vavriv at al., 2011).

4.2 Correction of geometric distortions in SAR images by resampling

In the considered time-domain SAR processing algorithm with the clutter-lock, each line of a SAR image in the range direction represents the ground scene on the Doppler centroid line determined by the current antenna beam orientation angles $\alpha(t)$ and $\beta(t)$ in the actual local coordinate system (see Fig. 6). Thus, the application of the clutter-lock under unstable flight conditions leads to geometric distortions in SAR images as illustrated in Fig. 7. Such geometric distortions can be corrected by resampling of the images from the radar native coordinates "slant range – azimuth" to a correct rectangular grid on the ground plane (X, Y) by taking into account the measured aircraft trajectory and the orientation of the synthetic aperture beams.

The resampling procedure consists of the following steps.

1. Define the reference flight line and the reference parameters, as well as the corresponding scene coordinate system for a given SAR image frame as it was described in Section 3.1.

2. Perform the resampling (interpolation) of the SAR image $SAR(X_A, R)$ from the slant range to the ground range in four steps:

    2.1. Calculate the coordinates of the image pixels $SAR(X_A, R)$ in the actual local coordinate system: $(x_{SAR}(X_A, R), y_{SAR}(X_A, R))$.

    2.2. Re-calculate the coordinates of the image pixels from the actual local coordinate system to the scene coordinate system according to (15), (16) and obtain the coordinates $(X_{SAR}(X_A, R), Y_{SAR}(X_A, R))$.

    2.3. Perform a one-dimensional interpolation of the SAR image line-by-line in the range direction from the uniform grid in the slant range to the uniform grid in the ground range. As the result, we obtain the image $SAR(X_A, Y)$.

    2.4. Find the coordinates $X_{SAR}(X_A, Y)$ of the image samples $SAR(X_A, Y)$ in the scene coordinate system from the coordinates $X_{SAR}(X_A, R)$ by the same one-dimensional interpolation in the range direction.

3. Perform the interpolation of the SAR image in the azimuth direction in the following two steps:

    3.1 Perform a joint sorting of the pairs of the range-interpolated image samples $SAR(X_A, Y)$ and their azimuth coordinates $X_{SAR}(X_A, Y)$ in the ascending order with respect to the X_A-coordinate. This step is required to correct significant forward-backward sweeps of the synthetic beam caused by motion errors.

    3.2 Perform a one-dimensional interpolation of the SAR image samples $SAR(X_A, Y)$ from the initial non-uniform grid of the along-track azimuth coordinate X_A to the uniform grid X. The result is the desired image $SAR(X, Y)$ in the ground scene coordinates.

The above described resampling algorithm is typically performed as a post-processing procedure.

4.3 Experimental results

The described in Sections 4.1 and 4.2 SAR processing approach has been implemented in the airborne RIAN-SAR-Ku system (Vavriv at al., 2006; Vavriv & Bezvesilniy, 2011a; Vavriv at al., 2011). The light-weight aircrafts Antonov AN-2 and Y-12 were used as the platform.

An example of a single-look SAR image built by the described SAR processing algorithm with the clutter-lock is shown in Fig. 10a. This is the SAR image before the correction of the geometric distortions by resampling. The image resolution is 3 m. The "forward-backward-

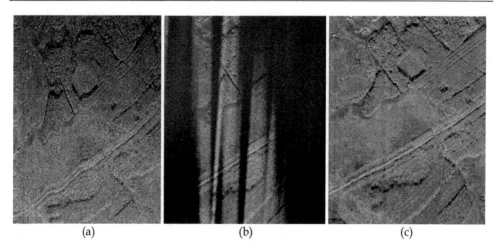

Fig. 10. Geometric distortions in a single-look SAR image built by using the clutter-lock (a), radiometric errors in the multi-look SAR image built without the clutter lock (b), the multi-look SAR image without errors after the resampling procedure (c).

forward" motion of the antenna beam leads to the evident distortions of the road lines and the contours of the forest areas in this image.

A 5-look SAR image formed without the clutter-lock is shown in Fig. 10b. The characteristic amplitude of the antenna beam orientation instabilities was larger than the 1-degree antenna beam width what resulted in significant radiometric errors. It should be noted that the proposed clutter-lock method based on the estimation of the antenna beam from the Doppler centroid measurements is efficient enough to avoid these radiometric errors in Fig. 10a.

The SAR images in Figs. 10a and 10b illustrates the dilemma of "geometric distortions vs. radiometric errors". Radiometric errors are removed due to the clutter-lock in Fig. 10a at the expense of geometric errors. And, vice versa, geometric errors are eliminated in Fig. 10b built without the clutter-lock, but at the cost of significant radiometric errors.

The application of the proposed resampling procedure resolves the dilemma, as it is illustrated in Fig. 10c. In this figure, both geometrical and radiometric errors are corrected.

5. Time-domain SAR processing with built-in geometric correction and multi-look radiometric correction

In this section, we describe a SAR processing approach, in which the correction of geometric distortions in SAR images is considered as the primary goal. We proposed (Bezvesilniy et al., 2010a; Bezvesilniy et al., 2010b; Bezvesilniy et al., 2010d; Bezvesilniy et al., 2011a) an algorithm called "built-in geometric correction" to control the synthetic beam direction so that the beams are pointed to the nodes of a correct rectangular grid on the ground plane. As the result, the SAR images are geometrically correct after the synthesis. The synthetic beams are obviously set to the nodes regardless of the real antenna beam orientation. The radiometric errors that arise in this case are corrected by a multi-look processing with extended number of looks.

5.1 Multi-look SAR processing on a single-look interval of synthesis

The multi-look processing in the time domain is usually performed by the coherent processing on sub-intervals of a long interval of the synthesis as described in Section 2.2. According to such approach, it is assumed that there are no significant uncompensated phase errors during the long time of the synthesis $T_{S\max}$ determined by (10). However, as a matter of fact, in order to achieve the desired azimuth resolution it is sufficient to perform the coherent processing on the short time interval T_S given by (9). This fact gives an alternative realization of the multi-look processing in the time domain, which is more preferable in the case of significant motion errors. The idea of the algorithm is to process the data collected during the short time of synthesis T_S with a set of different reference functions and migration curves to form the SAR look beams. We have called this approach "the multi-look processing on a single-look interval of synthesis". The proposed approach is illustrated in Fig. 11.

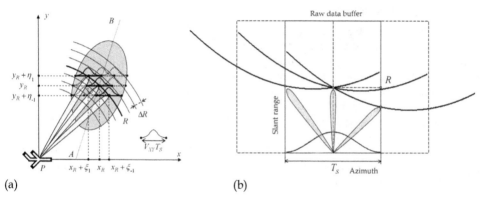

Fig. 11. The multi-look processing on a single-look interval of synthesis: (a) the antenna footprint consideration, (b) the raw data buffer consideration.

The reference functions of the different SAR looks should be built with the central frequencies (14) similar to the multi-look processing scheme in the frequency domain. The SAR look beam formed with the central frequency $\Delta F_C(R, n_L)$ is directed to some point $(x_R + \xi(R, n_L), y_R + \eta(R, n_L))$, which appears at the same slant range R at the centre of the short interval of the synthesis as illustrated in Fig. 11a. Let us derive formulas for these coordinates. The position of the point in the azimuth direction is related to its Doppler centroid (17), so we can write:

$$\Delta F_C(R, n_L) = \frac{2}{\lambda} \frac{(x_R + \xi(R, n_L))V_x - HV_z}{R}.$$ (21)

Substituting the expressions (14) and (17) into (21), we obtain:

$$\xi(R, n_L) = -n_L \frac{\lambda R}{2V_x} \frac{\Delta F_D}{2}.$$ (22)

Since the points, to which the synthetic beams of the SAR looks are directed, appear at the same slant range at the centre of the short interval of the synthesis, we can write:

$$x_R^2 + y_R^2 = (x_R + \xi(R, n_L))^2 + (y_R + \eta(R, n_L))^2 , \qquad (23)$$

and, finally,

$$\eta(R, n_L) = \sqrt{x_R^2 + y_R^2 - (x_R + \xi(R, n_L))^2} - y_R . \qquad (24)$$

Thus, in order to form the set of the synthetic beams of the different SAR looks on the short interval of synthesis for the slant range $R(n_L, X, Y)$, we should first calculate the points $(x_R + \xi(R, n_L), y_R + \eta(R, n_L))$ from (22) and (24), which correspond to the required central frequencies (14). Then, we should process the same raw data on the interval of the synthesis T_S with the appropriate range migration curves (5), the Doppler centroids (7) and the Doppler rates (8), by substituting the calculated coordinates $(x_R + \xi(R, n_L), y_R + \eta(R, n_L))$ instead of the coordinates (x_R, y_R) in these formulas.

The described approach to the multi-look processing has the following benefits. First, it is much easier to keep a low level of the phase errors on the short interval of synthesis, as compared to the long coherent processing time for all looks. Second, the orientation of the real antenna beam does not change significantly during the short processing time. This fact simplifies considerably the calculation of the orientation of all SAR look beams with respect to the real antenna beam for the subsequent radiometric correction. Third, a more accurate motion error compensation can be introduced in this processing scheme as compared to FFT-based algorithms. The compensation is performed based on the measured aircraft trajectory individually for each pixel of the SAR image accounting for both the range and the azimuth dependence of the phase and migration errors without any approximations. In other words, the accuracy of the motion error compensation is limited only by the accuracy of the trajectory measurements.

In the described approach, all SAR look beams are aimed at different points on the ground. It means that the obtained SAR look images are sampled on different grids. Therefore, the SAR look images should be first resampled to the same ground grid and only then they can be averaged to produce the multi-look image. Deviations of the aircraft trajectory introduce further complexity into the re-sampling process. We have proposed (Bezvesilniy et al., 2010a; Bezvesilniy et al., 2010b; Bezvesilniy et al., 2010d; Bezvesilniy et al., 2011a) an algorithm named "the built-in correction of geometric distortions" to solve this problem. This algorithm is described in the next section.

5.2 Built-in geometric correction

In order to avoid the interpolation steps in the above-described multi-look processing approach, the reference functions and the migration curves should be specially designed to point the multi-look SAR beams exactly to the nodes of a rectangular grid on the ground plane. The grid nodes to which the multi-look SAR beams should be pointed can be found as follows. The radar data are processed frame-by-frame forming a sequence of overlapped SAR images. For each frame, we define the reference flight line and the reference parameters before the synthesis of the aperture. The reference parameters of the data frame are used to calculate the Doppler centroid values $F_{DC}(R)$, the central Doppler frequencies $\Delta F_C(R, n_L)$ of the SAR looks, and the coordinates $(x_R^{ref} + \xi(R, n_L), y_R^{ref} + \eta(R, n_L))$ of the corresponding points on the ground in the reference local coordinate system. The found points are situated on the central frequency lines, which are similar to the Doppler centroid line AB in Fig. 1. The synthetic beams of the SAR looks should be pointed to the grid nodes, which are closest to the corresponding frequencies lines.

To point the SAR look beam to the found grid node, it is needed to recalculate the coordinates of this node from the reference local coordinate system to the actual local coordinate system by using (15), (16), taking into account the actual aircraft position and the orientation of the aircraft velocity vector. This recalculation is performed at each step of the synthesis. After that, the appropriate range migration curves (7), the Doppler centroids (8), and the Doppler rates (9) can be determined. Finally, the synthetic beam is formed to be directed to this node.

The proposed built-in geometric correction algorithm cannot be combined with the clutter-lock technique since the SAR beams do not follow the orientation of the real antenna beam. Therefore, the algorithm works well without an additional radiometric correction only for a wide-beam antenna and only for the central SAR looks. In order to use all possible SAR looks to form a multi-look SAR image without radiometric errors, we have proposed an effective radiometric correction technique based on multi-look processing with extended number of looks (Bezvesilniy et al., 2010c; Bezvesilniy et al., 2010d; Bezvesilniy et al., 2011b; Bezvesilniy et al., 2011c).

5.3 Radiometric correction by multi-look processing with extended number of looks

Let us denote an error-free SAR image to be obtained as $I(X, Y)$, where (X, Y) are the ground coordinates of the image pixels. This image is not corrupted by speckle noise and not distorted by radiometric errors. Whereas, a real SAR look image $I(n_L, X, Y)$ (n_L is the index of the SAR looks) is corrupted by speckle noise $S(n_L, X, Y)$ and distorted by radiometric errors $0 < R(n_L, X, Y) \le 1$ so that

$$I(n_L, X, Y) = I(X, Y) \cdot S(n_L, X, Y) \cdot R(n_L, X, Y) . \tag{25}$$

The speckle noise in a single-look SAR image (Oliver & Quegan, 1998) is a multiplicative noise with the exponential probability density function with the mean and the variance, correspondingly,

$$\mu\{S(n_L, X, Y)\} = 1 , \quad \sigma\{S(n_L, X, Y)\} = 1 . \tag{26}$$

The speckle noise is different for all SAR looks what is indicated here by the SAR look index n_L. The radiometric errors caused by instabilities of the antenna orientation can be considered as low-frequency multiplicative errors. The highest spatial frequencies of the radiometric error function $R(n_L, X, Y)$ are inversely proportional to the width of the real antenna footprint in the azimuth direction. Similar to the speckle noise, the radiometric errors are different for different SAR looks.

In order to compensate the radiometric errors, they should be estimated. For this purpose, we use a low-pass filtering $\mathbf{F}$ to measure the local brightness of the SAR images. This filter is designed to pass the radiometric errors and, at the same time, to suppress the speckle noise to some extent:

$$\mathbf{F}\{R(n_L, X, Y)\} \approx R(n_L, X, Y), \quad \mathbf{F}\{S(n_L, X, Y)\} \approx 1. \tag{27}$$

The application of this filter to the SAR look image (25) gives, approximately:

$$I_{LF}(n_L, X, Y) = \mathbf{F}\{I(n_L, X, Y)\} \approx I_{LF}(X, Y) \cdot R(n_L, X, Y). \tag{28}$$

Here $I_{LF}(X, Y)$ is the low-frequency component of the error-free SAR image to be reconstructed. The corresponding components of the real SAR looks $I_{LF}(n_L, X, Y)$ (28) contain information about the radiometric errors and they are almost not corrupted by speckle noise. These images can be used to compare radiometric errors on different SAR looks and, via such comparison, to estimate the radiometric errors. The idea of this empirical approach to the radiometric correction is based on the fact that one of many looks is pointed very closely to the centre of the real antenna beam. This look demonstrates the maximum power (brightness) among all looks, and this power is not distorted by radiometric errors.

Let us denote the number of looks to be summed up into the multi-look image as N_L^{pro}. This number of looks is slightly less than the number of the looks within the real antenna beam N_L since the orientation instabilities may corrupt the side looks considerably. By using the low-pass filter, it is possible to select the brightest (best-illuminated) parts of the scene among all extended SAR looks with the indexes $n_L = 1, ..., N_L^{ext}$ and compose only N_L^{pro} SAR looks (called the composite looks) for further processing. It is convenient to build the following sequence of the pairs of the composite looks and their low-frequency components:

$$\{I^{pro}(n_L^{pro}, X, Y), I_{LF}^{pro}(n_L^{pro}, X, Y)\}, \quad n_L^{pro} = 1, ..., N_L^{pro}. \tag{29}$$

This sequence is kept in the ascending order with respect to the brightness:

$$I_{LF}^{pro}(n_L^{pro}, X, Y) \le I_{LF}^{pro}(n_L^{pro} + 1, X, Y). \tag{30}$$

After processing of all the extended SAR looks, the brightest composite look is the look with the index $n_L^{pro} = N_L^{pro}$. These brightest values are obtained with the synthetic beams that are directed very closely to the centre of the real beam. Therefore, these brightness values are not distorted by the radiometric errors and give the estimate of the low-frequency component of the error-free SAR image to be reconstructed:

$$I_{LF}^{pro}(N_L^{pro}, X, Y) \approx I_{LF}(x, y). \tag{31}$$

This image can be used as the reference to estimate the radiometric error functions for all SAR looks:

$$R(n_L, X, Y) \approx \frac{I_{LF}(n_L, X, Y)}{I_{LF}^{pro}(N_L^{pro}, X, Y)}. \tag{32}$$

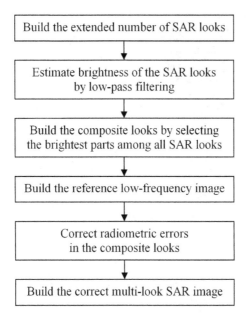

Fig. 12. The main steps of the multi-look radiometric correction algorithm.

By using the estimated radiometric error functions, radiometric errors for all SAR looks can be corrected before combining them into the multi-look SAR image. The main steps of the described algorithm are shown in Fig. 12.

If the navigation system is capable of measuring accurately the fast variations of the real antenna beam orientation, and if the real antenna pattern is known, the radiometric error functions (32) can be calculated directly from the relative orientation of the synthetic beam and the real antenna beam. This approach is more rigorous and accurate than the above-described empirical approach with the image brightness estimation. Nevertheless, with this approach, it is still necessary to build extended number of SAR looks, select the best parts of SAR images among all looks, and form the composite looks for multi-look processing.

5.4 Experimental results

The proposed approach has been used for post-processing of the radar data obtained with the RIAN-SAR-Ku and RIAN-SAR-X systems described in Section 7.

The performance of the built-in geometric correction is illustrated in Fig. 13. The SAR image shown in Fig. 13a is built by using the clutter-lock technique. One can see geometric distortions caused by instabilities of the antenna orientation. The undistorted SAR image shown in Fig. 13b is formed by using the algorithm with the built-in geometric correction. Both images have 3-m resolution and are built of 3 looks. The accuracy of the geometric correction is illustrated in Fig. 13c, where the SAR image built of 45 looks and formed by using the built-in geometric correction is imposed on the Google Map image of the scene.

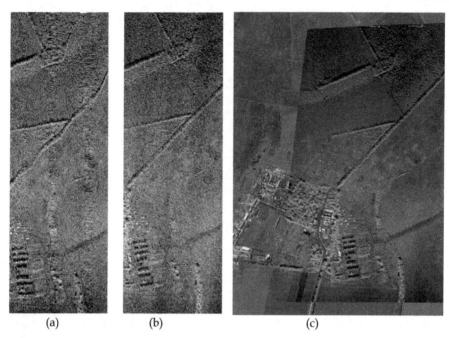

Fig. 13. Illustration of the geometric correction: (a) the 3-look SAR image built by using the clutter-lock technique, (b) the 3-look SAR image formed by using the built-in geometric correction, and (c) the 45-look SAR image formed by using the built-in geometric correction is imposed on the Google Maps image of the scene.

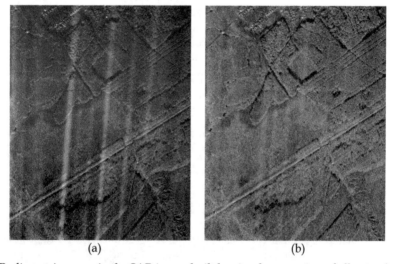

Fig. 14. Radiometric errors in the SAR image built by simple averaging of all extended SAR looks (a). SAR image formed of 5 composite SAR looks by using the proposed radiometric correction with extended number of looks.

The performance of the proposed radiometric correction by the multi-look processing with extended number of looks is illustrated in Fig. 14. The SAR image in Fig. 14a is built by simple averaging of all extended SAR looks. The image demonstrates good geometric accuracy; however the radiometric errors are presented. One can see dark and light strips in the image caused by the non-uniform illumination of the scene. The dark areas are due to the illumination for a short time when the real antenna footprint quickly moves to the neighbour areas of the scene. The light areas are correspondently illuminated for a longer time. The SAR image shown in Fig. 14b is built by using the proposed method of the multi-look radiometric correction with extended number of looks. The image is built of 5 composite SAR looks. One can see that the radiometric errors have been corrected successfully.

The obtained results prove that the described SAR processing approach can be effectively used for SAR systems installed on light-weight aircrafts with a non-stabilized antenna. An important advantage of the algorithm is that the produced SAR images are already geometrically correct at once after the synthesis, and there is no need in any additional interpolation. Another important advantage of the algorithm is the reduced requirements to the SAR navigation system. Although the aircraft velocity vector should be measured quite accurately to point the synthetic beams at the proper points on the ground, the aircraft trajectory should be measured and compensated with the high accuracy of a fraction of the radar wavelength only during the short time of the synthesis of one look. There is no need to keep so high accuracy of the trajectory measurement during the long time of the data acquisition for all looks.

6. Range-Doppler algorithm with the 1-st and 2-nd order motion compensation

The range-Doppler algorithm (RDA) is one of the most popular SAR processing algorithms. A high computational efficiency and a simplicity of the implementation are its main advantages. This algorithm belongs to the frame-based SAR processing algorithms, which use the FFT and work in the frequency domain. The motion compensation within the data frame is required. The SAR images are geometrically correct but they are originally produced in the radar coordinates "slant range – azimuth". Therefore, the ground mapping by an interpolation is required followed by stitching of the obtained image frames into the SAR image of the ground strip. Possible radiometric errors should be additionally corrected.

6.1 The 1-st and 2-nd order motion compensation

The geometry of the motion compensation problem is illustrated in Fig. 15. The point $A\,(0,0,H)$ indicates the expected position of the aircraft on the reference straight line trajectory. The point $A_E(\Delta x_E, \Delta y_E, H + \Delta z_E)$ corresponds to the actual position on the real trajectory. The slant range error for the synthetic beam directed to the point $P(x_R, y_R, 0)$ on the Doppler centroid line (1), (2) at the slant range R can be written as

$$\Delta R_E(x_R, y_R) = R_E(x_R, y_R) - R,\tag{33}$$

$$R_E(x_R, y_R) = \sqrt{(\Delta x_E - x_R)^2 + (\Delta y_E - y_R)^2 + (H + \Delta z_E)^2}.\tag{34}$$

These relations describe both the range migration errors and the corresponding phase errors

$$\varphi_E(x_R, y_R) = -\frac{4\pi}{\lambda}\Delta R_E(x_R, y_R) \tag{35}$$

caused by the trajectory deviations $\vec{r}_E = (\Delta x_E, \Delta y_E, \Delta z_E)$.

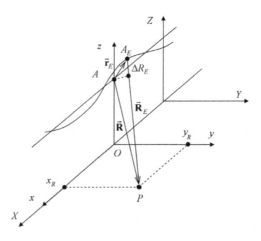

Fig. 15. Geometry of trajectory deviations.

In order to compensate the motion errors, we should correct the range migration errors (33), (34) by introducing an addition interpolation in the range direction and also correct the phase errors (35) in the azimuth direction. The corresponding correction should be performed individually for each pulse on the interval of the synthesis in the accordance with the current aircraft position error $\vec{r}_E = (\Delta x_E, \Delta y_E, \Delta z_E)$. The problem is that the range error ΔR_E depends not only on the slant range, but also on the direction to the point $P(x_R, y_R, 0)$. It means that the motion errors depend on both range and azimuth and are different for different points on the scene. In other words, the same radar pulses on two overlapped intervals of the synthesis should be compensated individually for the neighbour points in the azimuth direction. Such complete and accurate motion error compensation is possible only in those SAR processing algorithms, which allow the application of an individual reference function and range migration curve for each point of SAR image. It is possible, for example, in the time-domain SAR processing algorithms considered here. However, for the most SAR processing algorithms, including the range-Doppler algorithm, the dependence of the error ΔR_E on the azimuth must be disregarded and the range dependence is taken into account only.

The motion error correction should not interfere with the range and azimuth compression. Any range-dependent motion compensation can not be applied before the range compression of the received radar pulses. Otherwise, the range LFM waveform of the transmitted pulse will be distorted. Also, the range-dependent compensation cannot be applied before the range migration correction step of the SAR processing algorithm.

Otherwise, different corrections applied for the neighbour range bins will introduce phase errors in azimuth direction.

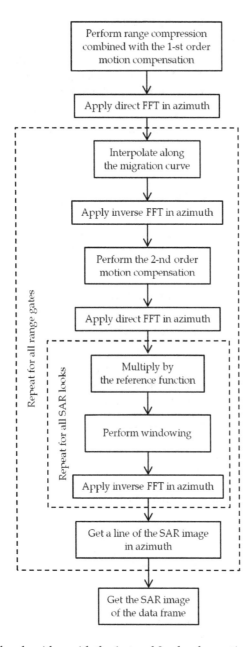

Fig. 16. Range-Doppler algorithm with the 1-st and 2-nd order motion compensation.

To cope with the above problems, the motion compensation procedure for the range-Doppler algorithm (and similar FFT-based algorithms) is usually divided on two steps (Franceschetti & Lanari, 1999):

1. First-order range-independent motion compensation,
2. Second-order range-dependent motion compensation.

The first-order motion compensation includes the range delay (33), (34) of the received pulses (with interpolation) and the phase compensation (35), which are calculated for some reference range, for example, for the centre range of the swath R_C:

$$\varphi_E^{(I)}(R_C,t) = \exp\left[-i\frac{4\pi}{\lambda}\Delta R_E^{(I)}(R_C,t)\right]. \tag{36}$$

Here t is the flight time. The first-order motion compensation can be incorporated into the range compression step but it should be performed before any processing step in the azimuth, in particular, before the range migration correction in the range-Doppler algorithm, as shown in Fig. 16.

The second-order range-dependent motion compensation is performed after the range compression and the range migration correction steps. It includes the phase compensation and may (or may not) include the following range interpolation step:

$$\Delta R_E^{(II)}(R, t) = \Delta R_E(R, t) - \Delta R_E^{(I)}(R_C, t). \tag{37}$$

$$\varphi_E^{(II)}(R, t) = \exp\left[-i\frac{4\pi}{\lambda}\Delta R_E^{(II)}(R, t)\right]. \tag{38}$$

Since the motion errors depend on time, it is needed to return from the range Doppler domain into the time domain by the inverse FFT, apply the corrections (37), (38), and come back into the range-Doppler domain by applying the direct FFT again, as shown in Fig. 16. After that, we can perform the azimuth compression.

After the compensation, the raw data seem like they are collected from the reference straight line trajectory, and the range-Doppler processing is performed by using the reference parameters of the data frame.

6.2 Problem of radiometric errors caused by motion compensation

The RDA performs processing of data blocks in the azimuth frequency domain assuming that the aircraft goes along a straight trajectory with a constant orientation during the time of the data frame accumulation. Therefore, instabilities of the antenna beam orientation within the data frame lead to radiometric errors in SAR images formed by the RDA.

After applying the above-described 1-st and 2-nd order motion compensation procedures, the corrected raw data demonstrate the range migration and phase behaviour as if the data were collected from the reference straight trajectory. However, the illumination of the scene by the real antenna is not changed, and radiometric errors are still presented.

Moreover, the application of the motion compensation can make the problem of radiometric errors even worse (Bezvesilniy et al., 2011c). After the motion compensation, the location of the antenna footprint on the ground should be described with respect to the position of the aircraft on the reference trajectory by the orientation angles α_{MoCo} and β_{MoCo} which are different from the angles α and β of the actual local coordinate system. The Doppler centroid values of the corrected radar data are apparently different from the Doppler centroid values before the motion compensation. For example, even if the antenna orientation is constant with respect to the actual local coordinate system, the orientation of the antenna beam can demonstrate variations with respect to the reference flight line. It means that the raw data with the constant Doppler centroid could demonstrate Doppler centroid variations after applying the motion compensation. This effect becomes more significant in the case of notably curved trajectories.

The problem of the correction of radiometric errors in SAR images formed by using the range-Doppler algorithm can be solved by the multi-look processing with extended number of looks as it was described in Section 5. It should be pointed out that the clutter-lock based on the estimation of the antenna beam orientation angles from the Doppler centroid measurements can be used together with the range-Doppler algorithm only to estimate the reference orientation angles for each data frame.

6.3 Experimental results

The range-Doppler algorithm with the 1-st and 2-nd order motion compensation procedures was implemented in the airborne RIAN-SAR-X system, what allows us to obtain multi-look SAR images in real time. The application of a wide-beam antenna enables avoiding radiometric errors in real time. An example of a 7-look SAR image with a 2-m resolution is given in Fig. 17. The application of the multi-look radiometric correction with the extended number of looks can be applied as a post-processing task to the recorded raw data.

Fig. 17. An example of a 7-look SAR image obtained with the X-band SAR system.

7. Practical Ku- and X-band SAR systems

In this section we describe the design and basic technical characteristics of the mentioned already the Ku- and X-band SAR systems developed and produced at the Institute of Radio Astronomy of the National Academy of Sciences of Ukraine (Vavriv at al., 2006; Vavriv & Bezvesilniy, 2011a; Vavriv at al., 2011). The SAR systems were designed to be deployed on a light-weight aircraft. The systems were successfully operated from AN-2 and Y-12 aircrafts.

7.1 Airborne system RIAN-SAR-Ku

The Ku-band SAR system RIAN-SAR-Ku Ukraine (Vavriv at al., 2006; Vavriv & Bezvesilniy, 2011a; Vavriv at al., 2011) operates in a strip-map mode producing single-look SAR images with a 3-meter resolution in real time. The radar can perform measurements at two linear polarizations. The system has also a Motion Target Indication (MTI) capability. Characteristics of the system hardware are listed in Table 1.

7.1.1 Hardware solutions

The radar transmitter is based on a traveling-wave tube power amplifier (TWT PA). The radar transmits long pulses with the duration of 5 μs. The binary phase codding technique is used for the pulse compression to achieve a 3-meter range resolution. The M-sequences of the length of 255 are used for phase codding. The transmitted pulse bandwidth is 50 MHz.

A high pulse repetition frequency (PRF) of 20 kHz is required in the system for detection of moving targets. The application of binary phase codding allows us to simplify dramatically the hardware realization of the range compression as compared to the well-known pulse compression technique of pulses with linear frequency modulation (LFM). It is critical to manage the range compression in real time at the high PRF of 20 kHz.

The pulse repetition frequency is adjusted continuously to keep the ratio of the aircraft velocity to the PRF constant. It means that the aircraft always flights the same distance during the pulse repetition period. Such approach is used to simplify the further SAR processing.

A sensitive receiver with the noise figure of 2.5 dB is used in the SAR. The system losses are 4.0 dB. The received data are sampled with two 12-bit ADCs at the sampling frequency of 100 MHz.

For the detection of moving targets, we used the following simple principle: All signals, which are detected outside of the Doppler spectrum of the ground echo, are assumed to be signals of moving targets. This approach calls for using of a narrow-beam antenna so that the Doppler spectrum from the ground is narrow. Therefore, a long slotted-waveguide antenna of the length of 1.8 m with a 1-degree beam has been used. The antenna is actually built of two separate antennas so that the SAR system can operate at two orthogonal linear polarizations.

The usage of such narrow-beam antennas is not common for airborne SAR systems. It imposes the following limitations on SAR imaging.

First, the azimuth resolution in the strip-map mode is limited by the half of the antenna length that is about 1 m for this system. If we degrade this resolution to 3 m, which is equal to the range resolution, it is possible to build only 5 half-overlapped SAR looks.

Parameter	RIAN-SAR-Ku	RIAN-SAR-X
Transmitter		
Transmitter type	TWT PA*	SSPA**
Operating frequency	Ku-band	X-band
Transmitted peak power	100 W	120 W
Pulse repetition frequency	5 – 20 kHz	3 – 5 kHz
Pulse repetition rate	< 200 Hz / (m/s)	Not used
Pulse compression technique	Binary phase codding (M-sequences)	Linear frequency modulation
Pulse bandwidth	50 MHz	100 MHz
Pulse duration	5.12 µs	5 – 16 µs
Receiver		
Receiver type	Analogue	Digital
Receiver bandwidth	100 MHz	100 MHz
Receiver noise figure	2.5 dB	2.0 dB
System losses	4.0 dB	1.5 dB
ADC sampling frequency	100 MHz	200 MHz
ADC capacity	12 bit	14 bit
Antenna		
Antenna type	Slotted-waveguide / Horn	Slotted-waveguide
Antenna beam width in azimuth	1° / 7°	10°
Antenna beam width in elevation	40° / 40°	40°
Antenna gain	30 dB / 21 dB	20 dB
Polarization	HH or VV / VV	VV
SAR Platform		
Aircraft flight velocity	30 – 80 m/s	30 – 80 m/s
Aircraft flight altitude	1000 – 5000 m	1000 – 5000 m
Aircrafts used	AN-2, Y-12	AN-2

* TWT PA is an acronym for a traveling-wave tube power amplifier.
** SSPA is an acronym for a solid-state power amplifier.

Table 1. Characteristics of the SAR hardware systems.

Second, the antenna beam orientation should be measured with a high accuracy of about 0.1° (that is 1/10th of the antenna beam width) to avoid radiometric errors in SAR images. The application of the antenna with a wider beam would simplify this requirement. An alternative horn antenna with a 7-degree beam was used to make the system capable of producing high-quality SAR images with many looks by processing of the recorded data.

7.1.2 Signal processing solutions

Radar data processing is performed with a special PCI-board equipped with a DSP and an FPGA. Characteristics of the SAR data processing system are given in Table 2.

Parameter	RIAN-SAR-Ku	RIAN-SAR-X
Range processing		
Range resolution	3 m	2 m
Range sampling interval	1.5 m	1.5 m
Number of range gates	1024	2048 (processed) / 4096 (raw)
Range swath width	1536 m	3072 m
Azimuth processing		
SAR processing algorithm	Time-domain convolution (stream-based)	Range-Doppler algorithm (frame-based)
Real-time motion error compensation (trajectory)	No	Yes, 1st- and 2nd-order MOCO
Clutter-lock*	Line-by-line	Frame-by-frame
Pre-filtering	Yes	Yes
Azimuth resolution	3.0 m	2.0 m
Number of looks (in real time)	1	1 – 15
Ground mapping of SAR images	Post-processing	In real time
Data recording		
Raw data	Range-compressed, 7-times decimated	Uncompressed, no decimation
Recorded raw data rate	12 MB/s	80 MB/s
Pre-filtered data, navigation data, SAR images, etc.	Yes	Yes
Other capabilities		
Detection and indication of moving targets	Yes	No

* Estimation of the antenna beam orientation angles from the backscattered radar data and updating the SAR reference functions.

Table 2. Characteristics of the SAR data processing systems.

The procedure of pre-filtering was implemented to reduce the high input data rate by a coherent accumulation and down-sampling of the data in azimuth from 20 kHz to about 100 Hz that is determined by the antenna beam width.

The time-domain convolution-based SAR processing algorithm with range migration correction by interpolation is implemented, as described in Section 4. This algorithm forms each pixel of the SAR image with a separate reference function and a migration curve. Therefore, the algorithm works well under unstable flight conditions. The algorithm is fast enough for the operation in real time, if the length of the convolution is not too long. With the narrow-beam antenna and the pre-filtering procedure, this requirement has been satisfied. The SAR processing system is able to build single-look SAR images with 3-meter resolution in real time. The number of range gates is 1024 resulting in 1536-meter range swath width.

In order to measure accurately the antenna orientation, the algorithm described in Section 4 for the estimation of the antenna orientation angles directly from Doppler frequencies of backscattered radar signals was introduced. The accuracy of the estimation is about 0.1°. The angles are updated about 10 times per second what is sufficient to track fast variations of the antenna beam orientation.

The estimated angles are used to realize the clutter-lock. The pre-filter and the SAR reference functions are updated rapidly to track variations of the antenna orientation, and thus to avoid radiometric errors in SAR images.

The radar system is able to record the range-compressed data at the data rate of about 12 MB/s to hard disk drives for post-processing. A 7-times decimation of the input data stream is used to reduce the data rate for recording. The pre-filtered radar data, the navigation data, and SAR images are recorded as well.

7.2 Airborne system RIAN-SAR-X

The X-band SAR system RIAN-SAR-X Ukraine (Vavriv & Bezvesilniy, 2011a; Vavriv at al., 2011) is capable of producing high-quality multi-look SAR images with a 2-meter resolution in real time. The system is designed to operate from light-weight aircraft platforms in side-looking or squinted strip-map modes. Characteristics of the radar hardware and the signal processing systems are listed in Tables 1 and 2.

7.2.1 Hardware solutions

The radar operates in the X-band. The transmitter is based on a modern solid-state power amplifier (SSPA). The peak transmitted power is 120 W. The radar transmits long pulses with a linear frequency modulation. A direct digital synthesizer (DDS) provides frequency sweeping. The pulse duration can be chosen from 5 to 16 μs. The transmitted pulse bandwidth is 100 MHz. It gives the range resolution of 2 m. The pulse repetition frequency is from 3 kHz to 5 kHz, and that guarantees an unambiguous data sampling in the azimuth.

A digital receiver technique has been implemented. The noise figure of the receiver is 2 dB. The system losses are 1.5 dB. We have used one 200-MHz ADC with a 14-bit capacity.

The radar uses a compact slotted-waveguide antenna with a 10-degree beam. The wide beam is used, first, to avoid radiometric errors during the formation of SAR images in real time, and, second, to enable building of high-quality SAR images with a large number of looks at a post-processing stage. The antenna is firmly mounted on the aircraft; however it can be installed either into a side-looking or a 40-degree-squinted position.

The SAR system is designed to be operated from a light-weight aircrafts. During test flights, the SAR system was successfully deployed on an AN-2 aircraft. The aircraft flight altitude could be from 1000 m to 5000 m, and the aircraft flight velocity is expected to be from 30 m/s to 80 m/s. The implemented SAR processing algorithms can operate beyond of these intervals of flight parameters with minor adjustments.

7.2.2 Signal processing solutions

A strip-map SAR processing is performed by using a frame-based range-Doppler algorithm with motion compensation, as described in Section 6. The SAR system is capable of producing SAR images with a 2-meter resolution formed of up to 15 looks in real time. A scheme with half-overlapped frames is implemented to provide continuous surveillance of the strip without gaps despite of possible motion instabilities.

The SAR navigation system is based on a simple GPS-receiver capable of measuring the aircraft position and the aircraft velocity vector. The measured position is used to link the obtained SAR images to ground maps, and also to know the flight altitude above the ground. The aircraft flight trajectory is integrated from the measured aircraft velocity with a sufficient accuracy to perform the motion compensation. The antenna beam orientation is estimated from Doppler frequencies of the backscattered radar signals. The pitch and yaw antenna orientation angles are used both for motion compensation and for the aperture synthesis. Such angle estimation is a kind of clutter-lock processing allowing to track variations of the antenna beam orientation by adjusting the SAR data processing algorithm from one radar data frame to another.

The signal processing system is divided on two main parts. The first part of the system performs: 1) range compression of LFM pulses combined with the 1st-order motion compensation, 2) calculation of Doppler centroid values for each range gate (by FFT in azimuth) and estimation of the antenna orientation angles, and 3) pre-filtering of the range-compressed data. This processing is performed in a special PCI board with a DSP and an FPGA.

The second part of the data processing system forms multi-look SAR images by using a range-Doppler algorithm with the 2nd-order motion compensation. This processing is performed on a PC with an Intel Quad Core CPU (the above-mentioned PCI board is installed on this PC). It gives a flexibility in setting the azimuth processing parameters and allows using the developed SAR system as a suitable test-bed for testing new modifications of various frame-based SAR algorithms.

Stitching of the obtained SAR images into a continuous strip map can be performed on a client PC (or a notebook), while viewing the data in real time or offline.

The SAR system is capable of recording the original uncompressed radar data on a solid-state drives organized in a RAID-0 array at the full pulse repetition rate up to 5 kHz. These data are stored together with the navigation data (original GPS measurements, integrated trajectories, estimated orientation angles, motion compensation curves, etc.), as well as the pre-filtered range-compressed data and the SAR images formed in real time. Recorded data are used further in our research and development activity on SAR systems.

8. Conclusion

The presented results indicate that some of the essential problems that limited the development of SAR systems for small aircrafts are solved. In particular, the problem of the antenna beam orientation evaluation has been solved by extracting this information from the Doppler shift of the radar echoes. This technique enables to use only a simple GPS receiver to provide a reliable SAR operation. Simultaneously, the problem of the correction of the geometrical distortions in SAR images has been solved via the introduction of a signal processing algorithm, which provides pointing multi-look SAR beams exactly to the nodes of a rectangular grid on the ground plane. The proposed multi-look processing algorithm with extended number of looks has demonstrated a high efficiency for the correction of radiometric errors. The suggested approaches have been successfully implemented in and tested with Ku- and X-band SAR systems deployed on small aircrafts. It should be pointed that these solutions are as well useful for SAR systems deployed on other platforms.

9. Acknowledgment

The authors would like to thank all of their colleagues at the Department of Microwave Electronics, Institute of Radio Astronomy of the National Academy of Sciences of Ukraine for their help and fruitful discussions. In particular, we indebt to Dr. V. V. Vynogradov, Dr. V. A. Volkov, Dr. S. V. Sosnytskiy, Mr. R. V. Kozhyn, Mr. S. S. Sekretarov, Mr. A. Kravtsov, Mr. A. Suvid, and Mr. I. Gorovyi for their essential contributions to the development of practical SAR systems.

10. References

Bamler, R. & Hartl, P. (1998). Synthetic aperture radar interferometry. *Inverse Problems*, Vol. 14, pp. R1-R54.

Bezvesilniy, O. O., Dukhopelnykova, I. V., Vynogradov, V. V., & Vavriv, D. M. (2006). Retrieving 3D relief from radar returns with single-antenna, strip-map airborne SAR. *Proceedings of the 6th European Conference on Synthetic Aperture Radar (EUSAR2006)*. 16-18 May 2006, Dresden, Germany. pp. 1-4. (CD-ROM Proceedings).

Bezvesilniy, O. O., Dukhopelnykova, I. V., Vynogradov, V. V. & Vavriv, D. M. (2007). Retrieving 3-D topography by using a single-antenna squint-mode airborne SAR. *IEEE Transactions on Geoscience and Remote Sensing*, Vol. 45, No. 11, pp. 3574-3582.

Bezvesilniy, O. O., Vynogradov, V. V. & Vavriv, D. M. (2008). High-accuracy Doppler measurements for airborne SAR applications. *Proceedings of the 5th European Radar Conference (EuRAD2008)*. 30-31 Oct. 2008, Amsterdam, The Netherlands. pp. 29-32.

Bezvesilniy, O. O., Gorovyi, I. M., Sosnytskiy, S. V., Vynogradov V. V. & Vavriv D. M. (2010a). Multi-look stripmap SAR processing algorithm with built-in correction of geometric distortions. *Proceedings of the 8th European Conference on Synthetic Aperture Radar (EUSAR2010)*. 7-10 June 2010, Aachen, Germany. pp. 712-715.

Bezvesilniy, O. O., Gorovyi I. M., Sosnytskiy, S. V., Vynogradov V.V. & Vavriv D.M. (2010b). Multi-look SAR processing with build-in geometric correction, *Proc. of the 11th Int. Radar Symposium (IRS-2010)*. June 16-18, Vilnius, Lithuania. Vol. 1. pp. 30-33.

Bezvesilniy, O. O., Gorovyi, I. M., Vynogradov V.V. & Vavriv D.M. (2010c). Correction of radiometric errors by multi-look processing with extended number of looks, *Proceedings of the 11th Int. Radar Symposium (IRS-2010)*. June 16-18, Vilnius, Lithuania. Vol. 1. pp. 26-29.

Bezvesilniy, O. O., Gorovyi, I. M., Sosnytskiy, S. V., Vynogradov, V. V. & Vavriv, D. M. (2010d). Improving SAR images: Built-in geometric and multi-look radiometric corrections. *Proceedings of the 7th European Radar Conference (EuRAD2010)*. 30 September - 1 October 2010, Paris, France. pp. 256-259.

Bezvesilniy, O. O., Gorovyi, I. M., Sosnytskiy, S. V., Vynogradov, V. V. & Vavriv, D. M. (2011a). SAR processing algorithm with built-in geometric correction. *Radio Physics and Radio Astronomy*, Vol. 16, No. 1, pp. 98-108.

Bezvesilniy, O. O., Gorovyi, I. M., Vynogradov, V. V. & Vavriv, D. M. (2011b). Multi-look radiometric correction of SAR images. *Radio Physics and Radio Astronomy*, Vol. 16, No. 4, pp. ???-??? (Accepted for publication).

Bezvesilniy, O. O., Gorovyi, I. M., Vynogradov, V. V. & Vavriv, D. M. (2011c). Range-Doppler algorithm with extended number of looks, *Proceedings of the 2011 Microwaves, Radar and Remote Sensing Symposium (MRRS-2011)*. August 25-27, Kiev, Ukraine. pp. 203–206.

Blacknell, D., Freeman, A., Quegan, S., Ward, I. A., Finley, I. P., Oliver, C. J., White, R. G. & J. W. Wood (1989). Geometric accuracy in airborne SAR images. *IEEE Transactions on Aerospace and Electronic Systems*, Vol. 25, No. 2, pp. 241-258.

Buckreuss, S. (1991). Motion errors in an airborne synthetic aperture radar system. *European Transactions on Telecommunications*, Vol. 2, No. 6, pp. 655–664.

Carrara, W. G., Goodman, R. S. & Majewski, R. M. (1995). *Spotlight Synthetic Aperture Radar: Signal Processing Algorithms*, Artech House, ISBN 0-89006-728-7.

Cumming, I. G. & Wong, F. H. (2005). *Digital Processing of Synthetic Aperture Radar Data: Algorithms and Implementation*, Artech House, ISBN 1-58053-058-3.

Franceschetti, G. & Lanari, R. (1999). *Synthetic Aperture Radar Processing*, CRC Press, ISBN 0-8493-7899-0.

Li, F.-K., Held, D. N., Curlander, J. C. & Wu, C. (1985). Doppler parameter estimation for spaceborne synthetic-aperture radars. *IEEE Transactions on Geoscience and Remote Sensing*, Vol. 23, No. 1, pp. 47-56.

Madsen, S. N. (1989). Estimating the Doppler centroid of SAR data. *IEEE Transactions on Aerospace and Electronic Systems*, Vol. 25, No. 2, pp. 134-140.

Moreira, A. (1991). Improved multilook techniques applied to SAR and SCANSAR imagery. *IEEE Transactions on Geoscience and Remote Sensing*, Vol. 29, No. 4, pp. 529-534.

Oliver, C. J. & Quegan, S. (1998). *Understanding Synthetic Aperture Radar Images*, Artech House, ISBN 0-89006-850-X.

Rosen, P. A., Hensley, S., Joughin, I. R., Li, F.-K., Madsen, S. N., Rodriguez, E. & Goldstein, R. M. (2000). Synthetic aperture radar interferometry. *Proceedings of the IEEE*, Vol. 88, No. 3, pp. 333-382.

Vavriv, D. M., Vynogradov, V. V., Volkov, V. A., Kozhyn, R. V., Bezvesilniy, O. O., Alekseenkov, S. V., Shevchenko, A. V., Belikov, A., Vasilevsky, M.P. & Zaikin D. I. (2006). Cost-effective airborne SAR. *Radio Physics and Radio Astronomy*, Vol. 11, No. 3, pp. 276-297.

Vavriv, D. M. & Bezvesilniy, O. O. (2011a). Developing SAR for small aircrafts in Ukraine. *Proceedings of the 2011 IEEE MTT-S International Microwave Symposium (IMS 2011)*. 5-10 June 2011, Baltimore, USA. pp. 1-4. (CD-ROM Proceedings).

Vavriv, D. M. & Bezvesilniy, O. O. (2011b). Potential of multi-look SAR processing. *Proceedings of the 5th Int. Conference on Recent Advances in Space Technologies (RAST 2011)*. 9-11 June 2011, Istanbul, Turkey. pp. 365-369.

Vavriv, D. M., Bezvesilniy, O. O., Kozhyn, R. V., Vynogradov, V. V., Volkov, V. A. & Sekretarov, S. S. (2011). SAR systems for light-weight aircrafts. *Proceedings of the 2011 Microwaves, Radar and Remote Sensing Symposium (MRRS-2011)*. August 25-27, Kiev, Ukraine. pp. 15-19.

Wehner, D.R. (1995). *High-Resolution Radar (2nd Ed.)*, Artech House, ISBN 0-89006-727-9.

Permissions

The contributors of this book come from diverse backgrounds, making this book a truly international effort. This book will bring forth new frontiers with its revolutionizing research information and detailed analysis of the nascent developments around the world.

We would like to thank Ramesh K. Agarwal, for lending his expertise to make the book truly unique. He has played a crucial role in the development of this book. Without his invaluable contribution this book wouldn't have been possible. He has made vital efforts to compile up to date information on the varied aspects of this subject to make this book a valuable addition to the collection of many professionals and students.

This book was conceptualized with the vision of imparting up-to-date information and advanced data in this field. To ensure the same, a matchless editorial board was set up. Every individual on the board went through rigorous rounds of assessment to prove their worth. After which they invested a large part of their time researching and compiling the most relevant data for our readers. Conferences and sessions were held from time to time between the editorial board and the contributing authors to present the data in the most comprehensible form. The editorial team has worked tirelessly to provide valuable and valid information to help people across the globe.

Every chapter published in this book has been scrutinized by our experts. Their significance has been extensively debated. The topics covered herein carry significant findings which will fuel the growth of the discipline. They may even be implemented as practical applications or may be referred to as a beginning point for another development. Chapters in this book were first published by InTech; hereby published with permission under the Creative Commons Attribution License or equivalent.

The editorial board has been involved in producing this book since its inception. They have spent rigorous hours researching and exploring the diverse topics which have resulted in the successful publishing of this book. They have passed on their knowledge of decades through this book. To expedite this challenging task, the publisher supported the team at every step. A small team of assistant editors was also appointed to further simplify the editing procedure and attain best results for the readers.

Our editorial team has been hand-picked from every corner of the world. Their multi-ethnicity adds dynamic inputs to the discussions which result in innovative outcomes. These outcomes are then further discussed with the researchers and contributors who give their valuable feedback and opinion regarding the same. The feedback is then collaborated with the researches and they are edited in a comprehensive manner to aid the understanding of the subject.

Apart from the editorial board, the designing team has also invested a significant amount of their time in understanding the subject and creating the most relevant covers. They scrutinized every image to scout for the most suitable representation of the subject and create an appropriate cover for the book.

The publishing team has been involved in this book since its early stages. They were actively engaged in every process, be it collecting the data, connecting with the contributors or procuring relevant information. The team has been an ardent support to the editorial, designing and production team. Their endless efforts to recruit the best for this project, has resulted in the accomplishment of this book. They are a veteran in the field of academics and their pool of knowledge is as vast as their experience in printing. Their expertise and guidance has proved useful at every step. Their uncompromising quality standards have made this book an exceptional effort. Their encouragement from time to time has been an inspiration for everyone.

The publisher and the editorial board hope that this book will prove to be a valuable piece of knowledge for researchers, students, practitioners and scholars across the globe.

List of Contributors

Ahmed Abdel-Hafez
Shaqra University, Kingdom of Saudi Arabia

Mohamad Hussien Taha
Hariri Canadian University, Lebanon

Ahmed Akl
1CNRS-LAAS, Université de Toulouse, France
UPS, INSA, INP, ISAE; LAAS, F-31077 Toulouse, France
College of Engineering, Arab Academy for Science, Technology, and Maritime Transport, Cairo, Egypt

Thierry Gayraud and Pascal Berthou
1CNRS-LAAS, Université de Toulouse, France
UPS, INSA, INP, ISAE; LAAS, F-31077 Toulouse, France

Nicolae Jula
Military Technical Academy of Bucharest, Romaina

Cepisca Costin
University Politehnica of Bucharest, Romania

Mariusz Wazny
Military University of Technology, Poland

Marco Leo
Consiglio Nazionale delle Ricerche- Istituto di Studi sui Sistemi, Intelligenti per l'Automazione, Italy

Ramesh K. Agarwal
Department of Mechanical Engineering and Materials Science, Washington University in St. Louis, St. Louis, MO, USA

Yu Gu
Research Assistant Professor, Mechanical and Aerospace Engineering (MAE) Department, West Virginia University, USA

Marcello Napolitano
Professor, West Virginia University, USA

Haiyang Chao
Post-Doctoral Research Fellow, MAE Dept., West Virginia University, USA

Francis Barchesky
M.S. Student, MAE Dept., West Virginia University, USA

Jason Gross
Ph.D., MAE Dept., West Virginia University, USA

A. Haddad, H. Griffiths and R.T. Waters
Cardiff University, UK

N.I. Petrov and G.N. Petrova
Istra, Moscow region, Russia

Oleksandr O. Bezvesilniy and Dmytro M. Vavriv
Institute of Radio Astronomy of the National Academy of Sciences of Ukraine, Ukraine

Printed in the USA
CPSIA information can be obtained
at www.ICGtesting.com
JSHW011445221024
72173JS00004B/949